Multiple Chemical Sensitivities

Addendum to
Biologic Markers in Immunotoxicology

Board on Environmental Studies and Toxicology

Commission on Life Sciences

National Research Council

NATIONAL ACADEMY PRESS
Washington, D.C. 1992

NATIONAL ACADEMY PRESS   2101 Constitution Ave., N.W.   Washington, D.C. 20418

The project was supported by the Environmental Protection Agency; the National Institute of Environmental Health Sciences; and the Comprehensive Environmental Response, Compensation, and Liability Act Trust Fund through cooperative agreement with the Agency for Toxic Substances and Disease Registry, U.S. Public Health Service, Department of Health and Human Services.

Library of Congress Catalog Card No. 92-80854
International Standard Book No. 0-309-04736-6

Additional copies of this report are available from the National Academy Press, 2101 Constitution Avenue, N.W., Washington, D.C. 20418

S582

Printed in the United States of America

TABLE OF CONTENTS

Introduction

Jonathan M. Samet and Devra Lee Davis

The diagnostic label of *multiple chemical sensitivity* is being applied increasingly, although definition of the phenomenon is elusive and its pathogenesis as a distinct entity is not confirmed. Multiple chemical sensitivity has become more widely known and increasingly controversial as more patients have received the label.

The emergence of multiple chemical sensitivity as a phenomenon that needs investigation coincides with recognition that myriad exposures to environmental agents are sustained in indoor and outdoor environments (National Research Council 1991; Samet, Marbury, and Spengler 1987, 1988). Regulatory activity has focused on outdoor air pollution, but children and adults in the United States and other developed countries spend most of their time indoors (National Research Council, 1991) and personal exposures to many air pollutants are thus determined largely by indoor pollutant concentrations. Monitoring studies conducted during the 1970s and 1980s showed that homes, public buildings, and nonindustrial workplaces can be contaminated by diverse gaseous and particulate pollutants that originate in unvented combustion, evaporation of volatile agents from various materials and solutions, grinding and abrasion, soil gas, and biologic sources (Samet, Marbury, and Spengler, 1987, 1988). Absorption of the pollutants can occur through the lungs, gastrointestinal tract and skin.

The concept that multiple chemical sensitivity is a distinct entity that is caused by responses to chemicals originated in the work of Randolph in the 1950s (American College of Physicians 1989, Ashford and Miller 1991). In the disease model proposed by Randolph, multiple chemical sensitivity consists of an inability to adapt to chemicals and the development of responsiveness to extremely low concentrations after sensitization (Randolph 1956); the model postulates multiple symptoms that reflect involvement of multiple organ systems. Randolph's pathogenic schema includes "adaptation." Symptoms can occur on exposure to chemicals or on withdrawal from exposure after an adaptive response has taken place. Randolph and others who apply this model of pathogenesis have used controlled exposures to establish the presence of multiple chemical sensitivity: patients are placed in environments judged to eliminate deleterious agents and then exposed to suspect chemicals.

Many of the physicians who apply that model are now referred to as clinical ecologists. Randolph and others founded the Society for Clinical Ecology in 1965 and it became the

American Academy of Environmental Medicine in 1984. This organization offers a definition of multiple chemical sensitivity which is also referred to as ecologic illness: "Ecologic illness is a poly-symptomatic, multi-system chronic disorder manifested by adverse reactions to environmental excitants, as they are modified by individual susceptibility in terms of specific adaptations. The excitants are present in air, water, drugs, and our habitats." Clinical ecology has been controversial, and committees of other specialty organizations have considered its diagnostic and therapeutic approaches to be inadequately supported by published studies (American Academy of Allergy and Immunology 1986; California Medical Association 1986; American College of Physicians 1989). Several recent reviews provide a more comprehensive background on multiple chemical sensitivity (Cullen 1987, Bascom 1989, Ashford and Miller 1991).

Although the literature addressing multiple chemical sensitivity is increasing, information on its frequency and natural history is lacking. Even defining multiple chemical sensitivity and developing criteria for diagnosis have proved difficult. Diverse pathogenic mechanisms have been postulated, but experimental models for testing them have not been established.

At the request of the Environmental Protection Agency, the National Research Council conducted a workshop to develop a research agenda to study the phenomenon of multiple chemical sensitivity. The workshop was convened with the overall objective of addressing the gaps in the scientific evidence on multiple chemical sensitivity. Workshop participants were multidisciplinary and included clinicians, immunologists, toxicologists, epidemiologists. psychiatrists, psychologists, and other involved in research or clinical activity relevant to the problem. The participants had an extensive range of experience and views on multiple chemical sensitivity. Three groups were formed to develop a research agenda. The subjects assigned were 1) case evaluation and criteria for diagnosis 2) mechanisms potentially underlying multiple chemical sensitivity and 3) epidemiologic approaches to investigating multiple chemical sensitivity.

This volume includes the papers prepared and presented by individual workshop participants; the papers have not undergone peer review. The papers offer a variety of views of the pathogenesis of multiple chemical sensitivity, its definition and diagnosis, and its management. They also illustrate the range of opinions as to the legitimacy of multiple chemical sensitivity as a unique clinical entity and the approaches used to treat it.

The paper by Lebowitz reviews key concepts of sensitization central to considering possible immunologic mechanisms of multiple chemical sensitivity. Lebowitz points to the distinction between sensitization (sensitized persons respond at lower doses than nonsensitized persons) and irritation. Burrell and Meggs also address the immune system and multiple chemical sensitivity; multiple chemical sensitivity is not seen as consistent with the specificity of immune responses, and authors point to immune mechanisms that might produce this phenomenon.

Bell emphasizes the need to use an integrated neuropsychiatric and biopsychosocial approach in considering multiple chemical sensitivity. Both she and Miller and Ashford suggest a possible role of the limbic system in triggering by odor. Experimental approaches to addressing mechanisms are described by Karol (an animal model) and by Meggs (a clinical research protocol).

Miller and Ashford address the difficult problem of defining multiple chemical sensitivity, offering this operational definition: "The patient with multiple chemical sensitivity can be discovered by removal from the suspected offending agents and by rechallenge, after an appropriate interval, under strictly controlled environmental conditions.

Causality is inferred by the clearing of symptoms with removal from the offending environment and recurrence of symptoms with specific challenge." In a separate paper, Miller and Ashford address the distinction between allergy and multiple chemical sensitivity.

Several papers describe patients in whom multiple chemical sensitivity has been diagnosed. Heuser and colleagues report on 135 patients, providing the results of serologic testing for various antibodies. Rea and colleagues review 16 years of experience with over 20,000 patients studied at the Environmental Health Center in Dallas. Fiedler and Kipen provide the results of a detailed neurobehavioral and psychologic assessment of a small number of patients who met rigid case criteria.

Welch describes aspects of occupational asthma, an entity caused by specific chemicals or exposures. DeHart summarizes the positions of four professional societies that question many of the concepts concerning multiple chemical sensitivity and clinical ecology.

Three working groups were formed and provided an approach to research that needs to be done in this area. The first working group provided lists of case criteria, candidate populations for investigation, and approaches for evaluating patients with multiple chemical sensitivity. The group advocated the use of an environmental control unit to test the adaptation-deadaptation hypothesis. The workshop participants agreed that the recommendations of this working group should have the highest priority in any research agenda.

The second group also proposed the use of controlled exposures of subjects in whom multiple chemical sensitivity is diagnosed and of control subjects. This group advocated the formulation of animal models, the investigation of tissues, and the development of a data base of chemicals, foods, and drugs reported to be associated with multiple chemical sensitivity.

The third group was concerned with epidemiologic approaches. A multicenter clinical assessment of patients was advocated for early attention. The results of the multicenter study would be used to develop methods for determining the prevalence of multiple chemical sensitivity. This group also suggested followup of populations with discrete and sudden chemical exposures.

This volume is published as an addendum to Biologic Markers in Immunotoxicology (NRC, 1992).

REFERENCES

American Academy of Allergy and Immunology, Executive Committee. 1986. Clinical ecology. J. Allergy Clin. Immunol. 78:269-270.

American College of Physicians. 1989. Clinical ecology. Ann. Intern. Med. 111:168-178.

Ashford, N.A., and C.S. Miller. 1991. Chemical Exposures. Low Levels and High Stakes. New York: Van Nostrand Reinhold. 214 pp.

Bascom, R. 1989. Chemical Hypersensitivity Syndrome Study. Baltimore: Department of Environment of the State of Maryland.

California Medical Association Scientific Board Task Force on Clinical Ecology. 1986. Clinical ecology—A critical appraisal. West. J. Med. 144:239-245.

Cullen, M.R. 1987. The worker with multiple chemical sensitivities: An overview. Occup. Med. State Art Rev. 2:655-661.

NRC (National Research Council). 1991. Human Exposure Assessment for Airborne Pollutants: Advances and Opportunities. Washington, D.C.: National Academy Press. 321 pp.

NRC (National Research Council). 1992. Biologic Markers in Immunotoxicology. Washington, D.C.: National Academy Press.

Randolph, T.G. 1956. The specific adaptation syndrome. J. Lab. Clin. Med. 48:934.

Samet, J.M., M.C. Marbury, and J.D. Spengler. 1987. Health effects and sources of indoor air pollution. Part I. Am. Rev. Respir. Dis. 136:1486-1508.

Samet, J.M., M.C. Marbury, and J.D. Spengler. 1988. Health effects and sources of indoor air pollution. Part II. Am. Rev. Respir. Dis. 137:221-242.

OVERVIEW

To develop a research agenda on multiple chemical sensitivity syndrome, participants were assigned to three multi-disciplinary groups. The groups were asked to focus on: 1) the design of a research protocol for clinical evaluation; 2) studies to evaluate relevant exposures and mechanisms; and 3) epidemiologic approaches. The specific charges were:

- *Development of a Working Definition for the Syndrome*
- *Evaluation of All Potential Causes*
- *Evaluation of Uncertainties Associated with the Data and Data Gaps*
- *Preparation of Report on Research Recommendations and Protocol*

The reports of the three working groups follow. Because consensus was achieved within each group on the substance and language of the recommendations, the reports are presented unedited, as accepted at the workshop.

Research Protocol for Clinical Evaluation

WORKING GROUP I

The group agreed that patients have been identified with a condition of multiple and often diverse symptoms that have been attributed to chemical agents in the environment. These patients may have recognized disease syndromes. However, symptomatology related to multiple chemicals is a distinct feature of these patients that is not classifiable by existing criteria used in conventional medical practice for psychiatric or physical illness. Thus, this feature cannot be uniquely coded in either DSM-III-R or ICD-9.

CASE CRITERIA

Criteria for the selection of cases for evaluation of multiple chemical sensitivity were discussed, and the committee agreed on criteria for research purposes as discussed below (definitions for other purposes were not addressed by this group):

Exposures and Mechanisms

WORKING GROUP II

There are clinical reports of a syndrome which has been described as follows: Chemicals (odorous and nonodorous VOC's, solvents, pesticides, etc.) cause a 'sensitization' (induction phase) in a subset of individuals. Upon subsequent exposure to lower concentrations, these individuals may respond in a polysymptomatic fashion (triggering). This 'sensitivity" may spread to other chemicals, foods or drugs. The underlying mechanism(s) is unknown at present but may include immunological, neurological, endocrinological, psychological, or other factors.

Currently, there is insufficient objective evidence for the entity called multiple chemical sensitivity caused by chemical exposures. Research is needed to test characteristics of the entity and define mechanisms that may be involved. Therefore the following research approaches are suggested.

HUMAN STUDIES

Following a comprehensive history, including environmental exposures, physical examination and appropriate laboratory testing, subjects diagnosed with MCS and control individuals, should be challenged by exposure to a mixture of offending agents determined historically. A double blind controlled procedure should be employed. The possible role of "adaptation" and "deadaptation" should be considered in the protocol. Endpoints for response should include immunologic, neurologic, endocrinologic, psychologic, social, etc. markers or measures. Dose-response relationships should be examined.

A second approach could be evaluation of individuals, over time, in their usual environment.

OTHER EXPERIMENTAL APPROACHES

Animal models should be developed that mimic the human syndrome. Exposures and parameters measured in the animal model should mimic those measured in humans wherever possible.

Tissues obtained by biopsy and necropsy from patients, animals and their controls should be evaluated for pathologic changes. In addition, functional and morphological changes in cells and tissues should be evaluated using *in vitro* techniques.

ADDITIONAL NEEDS

Attempts should be made to develop a data base of chemicals, foods and drugs, and associated signs and symptoms, which have been reported to be associated with MCS. Evaluation of this data base should be explored to identify possible linkages between exposure and mechanisms.

Epidemiology

WORKING GROUP III

The initial purpose and goal of epidemiologic research would be to establish the magnitude of the problem caused by the MCS phenomenon in the population and to characterize the cases sufficiently for further work. Full use of epidemiologic methods to study this issue is handicapped by the lack of a precise, agreed upon case definition. The lack of understanding of mechanisms and of the exposure patterns involved also restrict epidemiologic approaches at this time.

An important distinguishing feature of the syndrome is that the concentrations of contaminants to which the individual responds is reported to be orders of magnitude below the concentration to which the majority of people respond. Because the lack of an agreed upon case definition is a limiting factor in making progress, studies aimed at improving this definition should receive emphasis.

An early priority should be a multi-center clinical case-comparison study in occupational/environmental medicine clinics and other appropriate facilities. Patients who report responding with signs and symptoms to concentrations of environmental chemicals orders of magnitude below where the normal public responds would be enrolled. The study would take place at multiple sites using an agreed upon set of criteria. Clinical histories, batteries of tests and structured interviews should be used to define case characteristics and natural histories. The efficacy, specificity and sensitivity of different tests and instruments should be evaluated and validated between centers. The tests should, among others, assess psychologic-psychiatric conditions, immune function, neurotoxic reactions, etc. The study subjects would be compared with clinic controls (e.g., low back injuries) for the purpose of understanding case definition, the descriptive epidemiology of cases and general cross-sectional study.

The information from the multi-center study can then be used to construct a population-based study to determine the prevalence of the MCS entity and associated conditions. The provisional nature of the case definition suggests that it would be important for a broad set of symptom prevalences to be determined to allow flexible construction of a variety of definitions. Such descriptive studies could be started with review of the Health Interview and NHANES surveys conducted by the National Center for Health Statistics. It is possible that one or more of the population-based cohorts such as the Framingham,

11

Alameda County, Tecumseh, or Washington County cohorts could have an addition to follow-up surveys to probe for prevalence of conditions such as MCS.

Population based methods should also be used to determine the basic descriptive epidemiology of certain multi-organ conditions linked by some to MCS, such as SLE, scleroderma, MS and somatization disorder.

Epidemiology has an important role to play in supporting methods evaluation and development for test instruments used in clinical studies as well. Determination of normal ranges for new test modalities, establishing the sensitivity and specificity of screens and biomarkers, and the construction of survey instruments are examples where epidemiologic input is important and we recommend the inclusion of such expertise in these areas. The committee felt that the follow-up of a defined population subjected to a discrete and sudden chemical exposure (such as a spill or pesticide misapplication) would also be useful to assess the initiation of hypersensitivity to environmental chemicals and its natural history. It recommends that a protocol for evaluation and follow-up be developed and pre-tested and that resources be set aside for designated centers (e.g., environmental/occupational clinics) to respond rapidly in the event of an incident.

Key Concepts: Chemical Sensitization

Michael D. Lebowitz

INTRODUCTION

Sensitization to chemicals can be defined as changes in the organism, usually the immunochemical system, by exposure to a chemical such that further chemical exposure leads to recognition by the organism. Such recognition will lead to a response that is marked by a greater reaction at lower doses than what would be observed in non-sensitized individuals. This is usually called hypersensitivity (Turner-Warwick, 1978). Inhalation of the antigen/allergen in an individual previously sensitized leads to an allergic reaction, such as rhinitis or conjunctivitis. If the skin is sensitized, as in allergic contact dermatitis, then contact will cause an oedematous response and/or a rash. Pulmonary (airway) sensitization manifests itself through bronchial constriction or obstruction (Davies and Blainey, 1983; Hetzel and Clark, 1983; Ramsdale et al., 1985). Some chemicals can produce different types of "allergy". There are various known and hypothesized mechanisms for sensitization. There are also host susceptibility factors, including genetic predisposition, which will play a role in sensitization and in disease manifestation (Turner-Warwick, 1978; Gregg, 1983). There must be some differentiation also between irritation and sensitization (Burge et al., 1979; Newman-Taylor and Davies, 1981). Irritants do not have the immunochemical recognition signal. Irritation in a non-sensitized individual usually leads to a slower, less serious response, at higher doses. This does not prevent irritants from producing or enhancing inflammation, which can be prolonged, and which can produce hypersensitivity reactions (e.g., bronchial hyperreactivity) when stimulated (discussed below). There are about 10-20% of almost any population which show greater "sensitivity" to irritants, which is usually organ (e.g., eye) specific, by responding at lower doses (Weber, 1984; EPA, 1986). (Acute infectious diseases may produce some of the same symptoms also, which would require other means of differentiation.) Chemicals have different potencies as sensitizers, not entirely dependent on the mechanism of sensitization. For instance, platinum salts have a very high potency (i.e., up to 95% properly exposed become sensitized) while formaldehyde has a low potency (i.e., 1-5% properly exposed become sensitized) (Brooks, 1982; Newman-Taylor and Davies, 1981.).

MECHANISMS OF SENSITIZATION

Classic Sensitization

The B cell IgE mediates various immediate hypersensitivity states, including typical allergy to biological aero-allergens. IgE mediation is involved heavily in the pathogenesis of asthma (Hetzel and Clark, 1983; Turner-Warwick, 1978). Several chemicals, such as TDI, TMA, platinum salts, form protein bound haptens that act as antigens (Weill and Turner-Warwick, 1981; Brooks, 1982; Chan-Yeung, 1990; Newman-Taylor and Tee, 1990). Basically, IgE forms antibodies to certain antigens/allergens that are inhaled, absorbed, and/or ingested. This is the recognition (and sensitization) phenomena. Specific IgE then persists in the sensitized; it is an outcome that can be assayed. There are IgE receptor sites on basophils and mast cells. When an IgE antibody-antigen binding occurs, the IgE can attach to these receptors. This usually leads to degranulation of the cells and subsequent release of histamine, chemotactic factors for platelets and other active cells (e.g., eosinophils, neutrophils), and other active mediators (e.g., PG's/LTB). This starts a chain reaction (or cascade) of immuno-chemical phenomena which leads to the hypersensitivity response, and can lead to inflammation. (Ibid.) Anti-IgG can lead to basophil activation as well, as can the cells and some of the chemotactic factors (e.g., PAF) directly (Marone, 1989). (The mechanisms are discussed in more detail in the Immunology and Animal Models section.)   IgE is genetically linked, as are other immunoglobulins (Turner-Warwick, 1978; Gregg, 1983; Lebowitz et al., 1984; McGue et al., 1989). IgE also has an amplification-control cycle that probably involves some subsets of IgG and secretory IgA. IgG itself is important in sensitization and hypersensitivity (e.g., in TMA hypersensitivity). IgE mediated activity is associated also with T-lymphocyte activity in the lung, as found from broncho-alveolar lavage (BAL) studies (Gerblich et al., 1991). There can also be IgG blocking antibody that somehow prevents IgE mediated reactions. (Secretory IgA is the most abundant immunoglobulin in the lung, but its role in amplification-control of IgE is poorly understood. Turner-Warwick, 1978). Sensitization can occur in the lower respiratory tract, usually called hypersensitivity pneumonitis (or extrinsic allergic alveolitis), such as isocyanate disease. The nature of the protein-hapten deserves attention; T- and B-cells appear to have specific roles (Ibid.).   Mediators (such as lymphokines) and receptors, and of other factors require further elucidation. Secretory IgA, along with IgG, and T-cells (especially suppressor/cytotoxic CD8 cells) are usually involved in hypersensitivity pneumonitis (Schlueter, 1982; Patterson et al., 1990; Semenzato, 1991). Some of the other known IgE-mediated chemical hypersensitivities are platinum and nickel salts, cobalt, other isocyanates (in addition to TDI), other anhydrides (in addition to TMA, such as phthalic acid anhydride), and organic acids (such as plicatic acid from red cedar) which are essentially all low molecular weight compounds which form protein haptens (Brooks, 1982; Marone, 1989; Allegra et al., 1989; Nemery, 1990; Chan-Yeung, 1990). Based on skin reactions, chromium salts, and colophony flux probably have IgE mediated immediate hyper-sensitivity. Based on this, and sensitization through ingestion, there are other chemicals, such as formaldehyde and ethylene oxide, which may turn out to have some IgE mediated hypersensitivity (Rockel et al., 1989). The various metals/metal salts appear to have IgG involvement also (as shown by precipitin formation), as do isocyanates and anhydrides (Adams et al., 1988; Thurmond and Dean, 1988; Patterson et al., 1990). Some of the haptens formed will produce late phase reactions as well, related to influxes of eosinophils, PMN's, inflammatory mediators (op cit.). Non-atopics will produce IgE antibodies if antigen is administered with certain adjuvants (Marsh et al., 1972), though these responses are short-lived. This implies an IgE suppressive mechanism, which appears to be T cell mediated, and this suppression is probably due to CD8+ T cells (Diaz-Sanchez and Kemeny, 1990). Certain antigens can sensitize non-atopics as well (e.g., platinum salts, castor bean dust), which implies that they have IgE-potentiating properties. Such potentiation appears related to increased CD4+ T cells (Ibid.). Certain common chemicals (e.g., ozone) or mixes (e.g., diesel exhaust, including with HCHO) appear to be appropriate adjuvants (Meggs). Sometimes, elevated IgE and eosinophilia are associated with smoking (Burrows et al., 1989), and with acute and persistent respiratory disease, as in rubber workers involved in a thermoinjection process, where the suggested sensitization has not been related to one of the component chemicals (Bascom et al., 1990). Lymphocyte factors related to immunoglobulin activity also activate human basophils (Marone, 1989). One has to consider the triggering mechanism in predisposed individuals, such as specific triggers of inflammation (e.g., HCHO). Asthmatics and those with BHR, or those with such predispositions may be such susceptibles. It is claimed that those with autoimmune disorders (e.g., Lupus/SLE, rheumatoid arthritis-RA) may be predisposed also (Thrasher et al., 1989; Ashford). SLE and RA are associated with circulating immune complexes, which may play a role (discussed elsewhere) (Turner-Warwick, 1978). One also has to consider factors that affect interleukins which have such an impact in the chain of events after presentation with the antigen, including activation of basophils (e.g., by IL-3). IL-1 is a basic mediator of intercellular activity within and between immune system functions, including having a role in tissue homeostasis; IL-1 inhibitory activity (IHA) is reduced in interstitial lung disease (Nagai et al., 1991), and IL-1 is related to cell-mediated immunity as well.

Inflammation

Whatever the mechanism, inflammation (in airways, skin, possibly other organs) appears to be a key trait in many sensitization conditions (Reed, 1988; Bonini et al., 1989; Hogg et al., 1991). (Although it is discussed elsewhere as well, it is worth mentioning some major components of inflammation.) Epithelial release of mediators, including histamine and chemotactic factors, appears to be a major component of inflammation and a mechanism of importance in sensitization and subequent triggering by chemicals. The increased activity of mononuclear cells (including PAM's), polymorphonuclear cells (including neutrophils) and eosinophils are part of the basic mechanisms involving inflammation. The release of oxidants, such as hydrogen peroxide (H2O2), as with exposure to formaldehyde, and superoxide anion from these cells is a key contributor to damage and responses, occurs more in sensitized organisms, and can occur with im- mune complexes (Adams et al., 1988; Rossi et al., 1989; Allegra et al., 1989; Cerasoli et al., 1991). The lack of sufficient, effective antioxidants could be a mechanism of sensitivity (as in those with genetic defects, such as those who have little or no ceruloplasmin). It is interesting to note that only scarce airway inflammatory cell activity has been reported in at least some asthmatics between attacks (Saetta et al., 1989). Airway inflammation is probably the main cause of bronchial hyperreactivity by itself or by contributing to bronchial obstruction (although obstruction can occur by other mechanisms). The increase of airway or skin permeability, permitting processes that lead to mast cell or basophil degranulation, leads to mediator release. Mucosal edema, an aspect of inflammation, can increase epithelial permeability, alter smooth muscle dynamics, and stimulate neural pathways (Bucca and Rolla, 1989). The occurrence of immune complexes in situ is important also (as discussed above); basophils can be activated by C3a and C5a (Marone, 1989). Macrophages are major cells in acute and chronic inflammation, immune responses including the complement system and immune deficiencies, are involved in hematopoesis and coagulation, and promote proliferation of many other cells. They present antigen to T cells. They release plasminogen activator and classes of toxins such as neutral proteases. Exposure to lymphokines is part of macrophage activation. Derived factors can activate basophils directly (Turner-Warwick, 1978; Adams et al., 1988; Thurmond and Dean, 1988).

Other Mechanisms of Sensitization

There are several hypothesized mechanisms, each with preliminary experimental verification of its potential role. Epithelial cells in the mucosa of bronchi and intestine can express Ia antigens. Ia-bearing epithelium can present antigens to specific T-cells, to initiate the classic immune response (Allegra et al., 1989). "It is possible that epithelial cells, like all antigen- presenting cells, act as accessory cells in immune responses by producing IL-1, which is responsible for T-cell migration and proliferation." (Ibid.) Also, HLA-DR antigens from airway epithelial cells are seen in those with interstitial lung disease (Ibid.). HLA-DR is part of the MHC (major histocompatibility) locus, and MHC-peptide linkages have been shown to be specific for some compounds (op cit.). Further, there may be epithelial cell-surface receptors with potential protein (i.e., hapten) specificity, probably involving IgG binding and myosin, which would induce intracellular changes (op cit; DeMarzo et al., 1989).   Mast cells have receptors for a variety of antigens and histamine-releasing factors, in addition to anti-IgE and -IgG (Marone, 1989). It is unknown whether the "antigen" receptor would accept specific chemical-protein haptens. IgG, which has such high affinity for binding proteins, may play a role in this function. In rhinitis, and probably in asthma, there are specific chemical sensitive mast cells in the epithelium, such as formaldehyde-sensitive mast cells in the nasal epithelium, which are related to increased numbers of mast cell progenitors in airway mucosa (Otsuka et al., 1985, 1986, 1987; Barret and Metcalfe 1987). These epithelia can produce cytokines which promote mast cell growth in the presence of lymphocytes (Ibid.; Denburg et al., 1987). Local cells synthesized close to progenitors influences which type of mast cell develops, such that "derangements in the airway epithelium may result in a generation of factors which can induce a selective expansion of a mast cell sub-population." (Allegra et al., 1989). These mast cells, as found in bronchial secretions, BAL's and bronchial epithelium of asthmatics, are leaky and have the characteristics of the specific chemical-sensitive mast cells (as described above) (Allegra et al., 1989). The increased numbers of these specific mast cells can be activated in the airway, by provocation and possibly by other triggers, and cause bronchoconstriction only in those with these characteristics (Ibid.). The same exposure -epithelial mast cell response would be likely in the intestine as well, and possibly other locations where there are mast cells with their superficial location in epithelium. Epithelial cells are capable of activating mast cells also thru the release of specific mediators (e.g.,[15] HETE) (Marone, 1989).

The possibility of neuro-sensitizers has been mentioned as a possibility, involving the adrenergic, cholinergic and/or NANC (neither) system(s). Mast cell-nerve interactions are biologically significant as reviewed recently (Bienenstock et al., 1991). Endogenous biochemical compounds are important in the regulation of smooth muscle tone, which is more evident in sensitized individuals (such as asthmatics) (Barnes et al., 1984; Randem et al., 1987). The autonomic system appears to be involved as well (Nadel and Barnes, 1984; Postma et al.,1985). The adrenergic system may be more important in this regard (Barnes et al., 1984). Central parasympathetic efferent signals are important for the circadian rhythm of bronchial tone (Dreher and Koller, 1990). The endogenous factors may only differentiate sensitized individuals; this requires further exploration. Although there is no knowledge of the specificity of muscarinic receptors (in the cholinergic system) for specific haptens, changes in the patterns of muscarinic receptors in sensitized individuals requires further exploration (Paggiaro et al., 1989). Neuropepities are important in asthma (Barnes, 1991). Threshold doses of PGD2 and takykinins (neuropeptides in the NANC system) to produce acetycholine induced bronchospasm (a cholinergic phenomena) were modified in animals presensitized (with ovalbumen) (Omini et al., 1989), implying a different set and time-course of reactions in sensitized organisms. Psychological conditioning may cause most cell degranulation in the rat (Bienenstock et al., 1991). Other ANS involvement, including that related to the limbic system, and hypothalmic responses is discussed elsewhere (C.Miller, I.Bell).

Different Pathways

The hypersensitive also may use a different pathway that essentially does not exist in the general population (as stated also by Burrell, and partly discussed above), which may be genetically determined. This may relate to inflammation directly, or to BHR (which appears to be, at least as far as entrainment, an independent genetic trait). (Figure 1) Asthmatics have a similar distribution, but different immunoreactivity of myosin heavy chains in bronchial smooth muscle compared to normal subjects (De Marzo et al., 1989).

Irritant Mechanisms

Pulmonary alveolar macrophages are engaged in response to irritants. They can release chemotactic factors (including PAF, ECT, NCT) and other mediators after stimulation by irritants (op cit.). Lysosomal enzymes are released, specifically after phagocytosis and cell lysis. Other mononuclear cells may play a role also. Irritants also stimulate PMN's (neutrophils), which release proteases. Eosinophils release MBP endotoxin, and all release free radicals (see above). These cells are the hallmarks of inflammation and the biochemicals released increase it, even when the stimulant is an irritant. Irritants can also produce endothelial cell damage which results in arachadonic acid metabolism (and circulation of prostaglan- dins, leukotrienes). Increased permeability leads to exposure of mast cells or basophils, which can then be activated to release other mediators (Thurmond and Dean, 1988). However, the type and level of mediator activity is less under irritation stimuli. Stimulation of C-fibers lead to release of Substance P and other neuropeptides (NKA, NKB), including vasoactive intestinal peptide (VIP); these are part of the NANC (nonadrenergic noncholinergic) system. Such stimulation may lead to epithelial release of mediators, and mast cell degranulation (Marone, 1989). Stimulation of cholinergic mechanisms (the vagal afferent pathway) may be important, though there is no indication that muscarinic receptors would have a "memory" for specific irritants (see above). Thus, chemical irritant contact may produce symptoms of disease that mimic

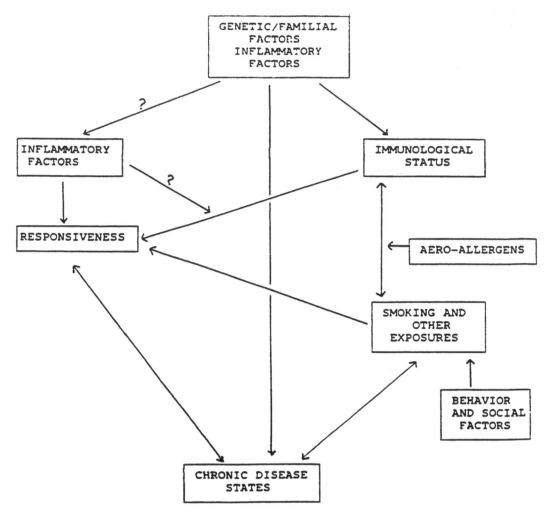

FIGURE 1   Hypothesized relationships.

immunological syndromes, but have a lack of antigen (Brooks, 1988; Thurmond and Dean, 1988). Protective responses, such as sneezing, coughing, and eye tearing, occur with stimulation of the epithelial irritant receptors. Irritants may even produce direct injury to epithelium. The "pseudoallery" or "reactive airway disease syndrome" usually requires exposure to relatively large amounts of the agent, usually occurs without a latency period after exposure (i.e., no "sensitization" period), and usually don't show reactivation of symptoms upon challenges with smaller amounts of the irritant or similar irritant (Ibid.). The irritant mechanisms last not much longer than due to the presence of the stimulus. Chronic inflammation can occur as well, leading to persistent symptoms, with high exposures to irritants; this may lead to reactive airway disease (Tarlo and Broder, 1989; Brooks, 1988). This form of chronic inflammation can result in increased susceptibility to other agents and to infection; implicated agents include chlorine, ozone, particles, SOx, NO2 (Gearhart and Schlesinger, 1986; Bylin et al., 1988; Wegman and Eisen, 1990). Examples of exposure agents that may produce such disease include acids, chlorine, phosgene, paints, calcium oxide, sulfur dioxide (Tarlo and Broder, 1989). Irritant stimuli may lead to adaptation or tolerance as well; ozone, NO2, H2S, phosgene and calcium oxide are agents that lead to reduced/attenuated responses (Wegman and Eisen, 1990). Irritant responses must be differentiated also from odor responses (Cain et al., 1987). The mechanisms of such tolerance or adaptation are not known. Certain "sensitizers" are pulmonary, dermal and sensory irritants as well, such as isocyanates, anhydrides, formaldehyde, ethylene diamine and paraphenylenediamine (Cain et al., 1987). Differentiation of responses is not always easy initially. After sensitization, stimulation can become non-specific as well, such that these and other irritants may produce similar responses. Even odors can produce non-specific responses in asthmatics (Shim and Williams, 1986).

FORMALDEHYDE

Formaldehyde is ubiquitous, especially in the indoor environment (Table 1). Formaldehyde is regarded as a upper respiratory irritant because it has a high solubility in water and is therefore captured by the wet mucus membrane. Retention of inhaled formaldehyde is shown to be almost 100% at the inhaled site in dogs (Egle, 1972). Malorney et al., (1965) showed that during an immediate following intravenous (IV) infusion of formaldehyde, the erythrocytes rapidly pick it up, leading to a very short-term increase in plama befor the erythrocytes oxidize it to formic acid. This is partly (1/6th) secruted in urine; the rest is oxidized to carbon dioxide and water. Formaldehyde (HCHO) irritant effects are presumed to be due to a nonspecific bronchial reflex response, which can increase flow resistance (Coffin and Stokinger, 1977). The effects of formaldehyde may be aggravated by interactions of other chemicals (Goldsmith and Friberg, 1977). The detection of the odor of formaldehyde in the air starts at about 0.05 ppm (0.06 mg/m3) (WHO/EURO, 1987) and is recognizable through most concentrations of exposure (Table 2). Exposure of a half hour to an hour is all that is needed to produce primary irritation; there is a dose response (Loomis, 1979: NRC 1981a). The lowest concentration causing pharyingitis is said to be 0.5 ppm (Stokinger and Coffin, 1968). Exposure to concentrations of formaldehyde gas of 0.1 ppm causes general irritation of the upper respiratory tract and other symptoms in children (Burdach and Wechselberg, 1980; NRC, 1981a). Concentrations from 1-5 ppm provoke coughing, constriction in the chest, and other symptoms (Loomis, 1979; NRC, 1981; Bardana, 1980; and Walker, 1964). Tolerance to eye and upper airway irritation may occur after 1 to 2 hours of exposure (NRC, 1981a; Blejer and Miller, 1966; Kerfoot and Mooney, 1975; Shipkovitz, 1968). However, even if tolerance develops, the irritation symptoms can return after 1 to 2 hour interruptions of exposure (Ibid; CPSC Ad Hoc Task Force, 1979). Ingestion usually leads injury to the larynx and trachea, gross effects on the GI tract, pneumonia, hardening of the lungs, hyperemia and edema of the lungs, and systemic damage. Increased mucolytic (decreased ciliary) activity has been shown related to formaldehyde exposure in animals in various studies (Carson et al., 1966; Cralley, 1942; Dalhamn, 1956; Hoffmann and Wynder, 1977; Kensler and Battista, 1963; Kotin, 1966). Decreased ciliary activity reduces the host defense mechanism to chemicals and micro-organisms, and may lead to increased cell wall permeability. Acute respiratory infections have been associated with formaldehyde exposure in children (Burdach and Wechselberg, 1980; NRC, 1981a; Tuthill, 1984). Trinkler (1968) found inflammatory changes in bronchial tracts of subjects who used formalin solution as a histological fixative.

Orringer and Mattern (1976) and Rockel et al., (1989) saw an increase frequence of acute and chronic bronchitis in dialyzed patients. There was a reactive, eosinophilic component.

TABLE 1

Contribution of Various Atmospheric Environments to Average Exposure[1]

Source	mg/d
1. Air	0.02
Ambient (10% 0f time)	
Indoor	
Home (65% of time)	
Conventional	0.5-2
Prefabricated (particle board)	1-10[3]
Workplace (25% of time)	
No occupational exposure[2]	0.2-0.8
With 1 mg/m[3] occupational exposure	4-5
ETS	0.1-1
2. Smoking (20 cigarettes/d)	1

[1] Contributions of food and water is low, so they are ignored here.
[2] Assuming the normal formaldehyde concentration is in conventional buildings
[3] Currently unusual.

Source: WHO/EURO 1987 Air Quality Guidelines for Europe (EURO Series #23) Copenhagen, 1987, pp 91-104.

Potential Mechanisms of Formaldehyde Induced Respiratory Sensitivity

Formaldehyde possesses the capacity to bind and alter protein constituents, such as: amino acids, proteins, nucleic acids, nucleosides, and nucleoproteins (Morin et al., 1964; Semin et al. 1974.; Goh el al., 1979; Bardana, 1980). Therefore, it is quite possible that this reactive chemical combines with respiratory tract proteins to form an immunoreactive, hapten-protein complex. Formaldehyde reacts directly with free amino groups, altering the

TABLE 2

Effects of Formaldehyde on Humans After Short-Term Exposure

Concentration of
formaldehyde
mg/m^3

Estimated median	Reported range	Effect
0.1	0.06 - 1.2	Odor threshold in 50 % of people (including repeated exposure)
0.5	0.1 - 1.9	Eye irritation threshold
0.6	0.1 - 3.1	Throat irritation threshold
3.1	2.5 - 3.7	Biting sensation in nose, eyes
5.6	5 - 6.2	Tolerable for 30 min (tearing)
17.8	12 - 25	Strong flow of tears, lasting for 1 hr
37.5	37 - 60	Danger to life, edema, inflammation, pneumonia
125	60 - 125	Death

Source: WHO/EURO 1987 Air Quality Guidelines for Europe (EURO Series #23) Copenhagen, 1987, pp 91-104.

character of the proteins, which could make the chemically altered constituents antigenic in themselves or can result in the formation of macro-molecular formaldehyde protein complexes and thus produce hyper- sensitivity in some persons (Bardana,1989; Loomis, 1979; Stokinger and Coffin, 1968). Supposedly, an IgE ISO RAST (Pharmacia, Sweden) was developed for formaldehyde and was positive for exposed dialysis patients (Rockel et al., 1989); others have evidently used formaldehyde conjugated to human serum albumen for immunoglobulin determinations (see below). Anti-N-like antibodies were found in dialysis patients using HCHO-treated cells (Sandler et al., 1979), which is evidence of type II auto-allergy (Consensus Workshop, 1984). Complement activation (via the alternate pathway) is another possible mechanism (Hakin et al., 1984; Rockel et al., 1989). C3a and C5a can activate basophils (Marone, 1989). More than one mechanism may be at work (Ibid.). Thrasher et al. (1990) report increased IL2 receptor cells and total B Cells (using monoclonal antibodies to LEU10) in those with long-term HCHO exposures (by report) and various presenting multiple organ system symptoms, compared to controls (with short-term exposure as chiropractic students. They also found increased titres (1:8 or greater) to B-cell

(IgG, IgM and IgE) isotypes to formaldehyde conjugated to human serum albumen; they indicate that these isotypes have proven positive in other investigations; increases in both IgG and IgE are hard to explain, and IgM increases haven't been found related to such immune disorders before (see above section, Turner-Warwick, 1978). Helper/Suppressor T-cell ratios don't appear different. They also reported increased titres (1:20) to auto antibodies (ASS, APC, ABB, AMIT and ANA). The meaning of significant increases in ASS only in mobil home residents, of APC and AMIT in variable percents in different groups, of ABB in office workers, and of ANA in both mobile home and office occupants are difficult to interpret. Increases in ANA hadn't been found in occupational asthma (Newman-Taylor and Davies, 1981; Newman-Taylor and Tee, 1990; op cit.). Other possible mechanisms have been discussed previously. It is quite likely that eosinophilia may be part of the immunological picture in formaldehyde-induced sensitivity (Popa et al., 1969; Hoy and Cestero, 1979; Nagornyi et al., 1979; Orringer and Mattern, 1976; Rockel et al., 1989). It has been shown that aldehyde pyrolysates from thermal degradation of polyethylene wrap is characterized by eosinophilia as well as by nocturnal attacks of dyspnea, wheezing and cough (Skerfving et al., 1980). There are an increasing number of studies which suggests formaldehyde can induce an inflammatory bronchitis reactivity airway disease. Particulates will carry formaldehyde gas (LaBelle et al., 1955; Patterson et al., 1990; Takhirov, 1974). Thus formaldehyde might have its affects in tracheobronchial or alveolar regions. Thus, particulates or aerosols have synergistic effects (Krzyzanowski et al., 1990), or the respiratory response is potentiated with simultaneous exposure to particulates as well as to other irritative chemicals (NRC, 1981b), or the aerosolized HCHO produces both sensitization and hyperresponsiveness (Hendrick and Lane, 1975, 1977; Frigas et al., 1982; Burge et al., 1985). Ostapovich (1975) studied sensitization of formaldehyde (2, 7, 15 mg/m3) in rats and guinea pigs injected with Freund's adjuvant prior to priming in order to stimulate antibody formation. Intermittent exposure, to concentration of 7 mg/m3 were given. Sensitization was induced by unequivocal increases in leukocyte agglutination, agar gel precipitation and passive hemaglutination, decreases in all but cholinesterase activity, from concentrations of 2 and 7 mg/m3. The time required for leukocytosis was shorter in rats than in guinea pigs. Zaeva (1968) summarized similar findings in animals. Nagornyi et al. (1979) found similar results, and eosinophilia, in guinea pigs with chronic exposures to 0.5 mg/m3. Sensitization was enhanced by intermittent inhalation. One mechanism of sensitization measured by leukocyte activity was formulated by Jerne, whose modified technique was utilized in this above study. This mechanism states that somatic generation of large number of genes expressed in leukocytes initially are directed against the animal's own histocompatibility antigens, which in the process of reacting with the cells that can form antibodies against them, cause these cells to mutate early, giving rise to many clones of antibody-forming cells with new and different specificities (Rose et al., 1973). Other in vitro and animal studies (Adams et al., 1988; Damiani, 1989) have shown that formaldehyde (like SO2 and nickel salts) increase macrophage phagocytosis and increase hydrogen peroxide levels. The latter promotes lipid metabolism, increases inflammation, and decreases host resistance; these studies have shown little direct effect of HCHO on B cells/immuno-globulins. Activated macrophages are associated with increased IL-1 and subsequent enhanced inflammatory processes (also by secretion of lytic enzymatic chemo attractants). (Increased IL-1 also stimulates T cell migration.) Phagolysomes degrade foreign molecules and resulting peptides associated with class I or II major histocompability complex (MHC) glycoproteins (which secondarily are presented as antigens to T cells (Damiani, 1989). Allegra et al. (1989) indicate that HCHO-sensitized most cells in the epithelium are associated with increased progenitors in airway mucosa and thus symptom severity. Mast cell promotion and differentiation in vitro occur by cytokines produced by such individual's epitherlium (with or without the presence of lymphokines) (op cit.)

Evidence of Asthma in Humans

The first case of formalin asthma was in an occupational setting and was published by Vaughan in 1939. Large numbers of European case studies reported on formaldehyde-induced asthma, claiming that an allergic mechanism was responsible, but immunological data is not presented usually (Vaughan, 1939; Coulant, 1961; Gervais, 1966; Tara, 1966; Nova and Touraine, 1956; Paliard et al., 1949; Sakula, 1975). Baur and Fruhmann (1979) showed an IgG increase in one case. In formaldehyde resin makers and other formaldehyde workers, formaldehyde, including formalin (40% formaldhyde by volume, an aqueous solution of formaldehyde) will produce primary irritational or allergic hypersensitivity. Further, formaldehyde reaction products, for example, dimethylurea, can be sensitizers (Milby et al., 1964). These and other studies (Hendrick and Lane, 1975, 1977; Popa et al., 1969; Kuzmenski et al., 1975; Kratchovil, 1971; Spassowski, 1976; led an NAS Committee (1981b) to conclude that formaldehyde has been shown to cause bronchial asthma in humans. The Committee stated that asthmatic attacks related to exposures to formaldehyde at low concentrations are due specifically to formaldehyde sensitization in some cases; controlled inhalation studies with formaldehyde are positive in these instances (Hendrick and Lane, 1975, 1977; Frigas et al., 1982; Burge et al., 1985). Nordman et al. (1985) found that only 12 of 230 persons with explicit suspicion of formaldehyde asthma actually reacted to formaldhyde gas broncho-provocation tests. They had significant dose-response reactions between $0.4 - 2.5$ mg/m3 HCHO. Of 218 who didn't, 71 reacted to histamine broncho-provocation. Cockroft et al. (1982) reported two cases of occupational asthma caused by urea formaldehyde particle board, with positive histamine bronchial challenge tests, but no increased RAST specific IgE antibodies directed against formaldehyde human serum albumin conjugate. A third asthmatic didn't have a response with exposure, indicating (to them) a specific sensitization to the agent. Formaldehyde also seems to act as a direct airway irritant in persons who have bronchial asthmatic attacks from other causes (op cit.) Hakim et al. (1984) also reported acute asthma-like symptoms of chest tightness and dyspnea in dialysis patients when formaldehyde-treated membranes were used. The symptoms were correlated with complement activation (C3a). Asthma-like symptoms and bronchospasm also were reported with similar exposure by Rockel et al.(1989), associated with increased eosinophilia, IgE, elevations, and complement activation (C3a, C5a). HCHO conjugated specific IgE (RAST) elevations were found in some; other agents were found to have similar effects (ethylene oxide, phthalates, isocyanate). Airway Obstructive Disease due to formaldehyde has been shown in various studies also (Yefremov, 1970; Kerfoot and Mooney, 1975; Schoenberg and Mitchell, 1975; Kilburn et al., 1985; Krzyzanowski et al., 1990). Most recently, a community population study (Krzyzanowski et al., 1990) showed higher rates of asthma and chronic bronchitis in children related to monitored formaldehyde exposure, with a synergistic contribution of ETS. Formaldehyde-induced pesistent bronchial responsiveness by itself occurred primarily in the lower SES group (Quackenboss et al., 1989a). The monitoring used a modified LBL method with no temperature or humidity interferences (Quackenboss et al., 1989b). They also found acute changes in children, especially asthmatics, who were specifically exposed overnight in their bedrooms; morning PEF had decreased significantly with a demonstrable exposure-response relationship.

Skin Sensitization

Skin contact with formaldehyde has been reported to cause a variety of acutaneoud problems in humans, including irritation, allergic contact dermatitis (Type I allergy), and urticaria (Type IV allergy) (Odom and Maibach, 1977; Roth, 1969; Sneddon, 1968; NRC, 1981a; Consenses Workshop, 1984). Dermal exposure to formaldehyde may cause an acute and inflammatory reaction of the affected region concentrations of 1 to 3 ppm over minutes to hours produces irritant effects which may be tolerated by normal adults, but produces noticeable skin irritation from direct exposure (Pirila and Kilipo, 1949; Roth, 1969).

Current Levels of Knowledge

It is commonly accepted there is reasonable good knowledge of expsoure assessment for formaldehyde, including: sources, monitoring instrumentation, and distribution of exposures (WHO/EURO, 1986, 1987, 1991) as represnted by Tables 1, 3 and 4. The suggestions of the Consensus Workshop (1984) for future research are pertinent still. Knowledge levels are reasonable for certain effects: odor, sensory and mucosal irritation. More knowledge is needed about airway effects, including sensitization (Ibid.) and neurological effects (Kilburn et al., 1985). It is estimated that between 1 and 6% of individuals in the US are probably sensitive to low concentrations of formaldehyde (Moschandreas et al., 1980). Means of control have included technical and regulatory procedures (WHO/EURO, 1991); educational control methods have been recommended also.

TABLE 3

Formaldehyde Sources With Usual Range of Related Concentrations Indoors	
Primary: Particle/ Press Board,	0.1 -2.0 mg/m3
Plywood (resin), Fabric	
Secondary: ETS	5- 50 Mg/ m^3
Wood Burning	
Infiltration: Vehicles in garages	?
From outdoors	?*

* Expected to increase with oxygenated fuels, especially methanol.

TABLE 4

Concentrations, Distribution of Exposures and Exposure-Response Relationships

Concentration (mg/m^3)	People with exposure		Levels of concern	
Range (mg/m^3)	Low	High	Limited	Definite
	most	few	<0.03	0.10
0-20-0.1-2.0				

Source: WHO/EURO 1991 (in press)
  The European Short-Term Air Quality Guideline (WHO/EURO, 1987)

REFERENCES

Adams, D. O., J. G. Lewis, and J. H. Dean. 1988. Activation of mononuclear phagocytes by xenobiotics of environmental concern: analysis and host effects. Pp. 351-369 in Toxicology of the Lung, D. E. Gardner, J. D. Crapo, and E. J. Massaro, eds. New York: Raven Press.

Allegra, L., L. M. Fabbri, G. Picotti, and S. Mattoli. 1989. Bronchial epithelium and asthma. Eur Respir J 2(6 Suppl.):460S-468S.

Bardana, Jr., E. J. 1980. Formaldehyde: hypersensitivity and irritant eactions at work and in the home. Immunol and Allergy Practice 2(3):11-23.

Barnes, P. J. 1991. Neuropeptides and asthma. Am Rev Respir Dis 143:S28-S32.

Barnes, P. J., G. A. Fitzgerald, M. Brown, and C. T. Dollery. 1984. Nocturnal asthma and changes in circulating epinephrine, histamine and cortisol. N Engl J Med 303:263-267.

Barrett, K. E., and D. D. Metcalfe. 1987. Heterogeneity of mast cells in the tissues of the respiratory tract and other organ systems. Am Rev Respir Dis 135:1190-1195.

Bascom, R., M. E. Baser, R. J. Thomas, J. F. Fisher, W. N. Yang, and J. H. Baker. 1990. Elevated serum IgE, eosinophilia, and lung function in rubber workers. 45(1):15.

Baur, X., and G. Fruhmann. 1979. Bronchial asthma of allergic or irritative origin as an occupational disease. Prax Klin Pneumol 1(Suppl. 33):317-22.

Bienenstock, J., G. Macqueen, P. Sestini, J. S. Marshall, R. H. Stead, and M. H. Perdue. 1991. Inflammatory cell mechanisms. Mast cell/nerve interactions in vitro and in vivo. Am Rev Respir Dis 143:S55-S58.

Blejer, H. P., and B. H. Miller. 1966. Occupational health report of formaldehyde concentrations and effects on workers at the Bayly Manufacturing Co., Visalia. Pp. 6 in Study Report No. S-1806. Los Angeles: State of California Health and Welfare Agency, Dept. of Public Health, Bureau of Occupational Health.

Bonini, Se., E. Adriani, St. Bonini, G. De Petrillo, G. G. Loggi, A. Adabbo, M. Pani, and F. Balsano. 1989. Allergy and asthma: distinguishing the causes from the triggers. An eye model for the study of inflammatory events following allergen challenge. Eur Respir J 2(6 Suppl.):497S-501S.

Brooks, S.M. 1982. The evaluation of occupational airways disease in the laboratory and workplace. J Allergy Clin Immunol 70(1):56-66.

Bucca, C., and G. Rolla. 1989. Mucosal oedema and airway hyperreactivity. Eur Respir J 2(6 Suppl.):520S-522S.
Burdach, S. and K. Weschselberg. 1980. Damage to health at school. Complaints due to the use of materials which emit formaldehyde in school buildings [German]. Forstschritte

der Medizin 98(11):379-384.

Burge, P. S., M. G. Harries, W. K. Lam, I. M. O'Brien, and P. A. Patchet. 1985. Occupational asthma due to formaldehyde. Thorax 40:255-260.

Burge, P. S., I. M. O'Brien, and M. G. Haries. 1979. Peak flow records in the diagnosis of occupational asthma due to colophony. Thorax 34:308-312.

Burrows, B., F. D. Martinez, M. Halonen, R. A. Barbee, and M. G. Cline. 1989. Association of asthma with serum IgE levels and skin-test reactivity to allergens. New Eng J Med 320:271-277.

Bylin, G., G. Hedenstierna, T. Lindvall, and B. Sundin. 1988. Ambient nitrogen dioxide concentrations increase bronchial responsiveness in subjects with mild asthma. Eur Respir J 1:606-612.

Cain, W. S., T. Tosun, L. C. See, and B. Leaderer. 1987. Environmental tobacco smoke: sensory reactions of occupants. Atmospheric Environment 21(2):347-353. Carson, S., R. Goldhamer, and R. Carpenter. 1966. Mucus transport in the respiratory tract. Am Rev Respir Dis 93 (Supl):86-92.

Cerasoli, Jr., F., J. Tocker, and W. M. Selig. 1991. Airway eosinophils from actively sensitized guinea pigs exhibit enhanced superoxide anion release in response to antigen challenge. Am J Respir Cell Mol Biol 4:355-363.

Chan-Yeung, M. 1990. Occupational asthma. Chest 98(5 Suppl):148S-161S.

Coffin, D. L., and H. E. Stokinger. 1977. Biological effects of air pollutants. Pp. 231-360 in Air Pollution. Vol. II The Effects of Air Pollution, A. C. Stern, ed. New York: Academic Press, Inc. Consensus Workshop. 1984. Report on the Consensus Workshop on Formaldehyde. Environ Health Perspect 58:323-381.

Coulant, M. P., and G. Lopes. 1961. L'allergie au formal et a ses derives. Archives des maladies professionelles, de medicine du travail et de securite sociale (Paris) 22:769.

Cralley, L. B. 1942. The effect of irritant gases upon the rate of ciliary activity. J Ind Hyg Tox 24:193-198.

Cockcroft, D. W., V. H. Hoeppner, and J.Dolovich. 1982. Occupational asthma caused by cedar urea formaldehyde particle board. Chest 1:49-53.

CPSC Ad Hoc Task Force. 1979. Formaldehyde. Washington, D.C.

Dalhamn, T. 1956. Mucous flow and ciliary activity in the trachea of healthy rats and rats exposed to respiraory irritant gases. Acta Physiologica Scandinavica 36(Suppl. 123):5-161.

Damiani, G. 1989. The role of macrophages and lymphocytes in bronchial asthma:

perspectives for new research models. Eur Respir J 6 (Suppl.):469S-472S.

Davies, R. J., and A. D. Blainey. 1983. Occupational Asthma. Pp. 205-213; 228-230 in Asthma, T. J. H. Clark and S. Godfred, eds. London: Chapman and Hall.

De Marzo, N., P. Di Blasi, P. Boschetto, C. E. Mapp, S. Sartore, G. Picotti, and L. M. Fabbri. 1989. Airway smooth muscle biochemistry and asthma. Eur Respir J 2(5 Suppl.):473S-476S.

Denburg, J. A., M. Ohnisi, J. Ruhno, J. Bienenstock, and J.Dolovich. 1987. The effects of basophil/mast and eosinophil cell colony stimulating activities derived from human nasal polyp epithelial scrapings are T cell-dependent. J Allergy Clin Immunol 79(Suppl.), 167A.

Diaz-Sanchez, D., and D. M. Kemeny. 1990. The sensitivity of rat CD8+ and CD4+ T cells to ricin in vivo and in vitro and their relationship to IgE regulation. Immunology 69:71-77.

Dreher, D., and E. A. Koller. 1990. Circadian rhythms of specific airway conductance and bronchial reactivity to histamine: the effects of parasympathetic blockade. Eur Respir J 3:414-420.

Egle, J. L. 1972. Retention of inhaled formaldehyde, propionaldehyde and acreolein in the lung. Arch Environ Health 25:119-124.

EPA. August 1986. Evaluation of Integrated Health Effects Data for Ozone and Photochemical Oxidants. Chapter 13 in Air Quality Criteria Document for Ozone and Other Photchemical Oxidants, EPA-600/8--84-020eF, Washington, D.C.

Frigas, E., W. V. Filley, and C. E. Reed. 1982. UFFI dust. Non-specific irritant only? Chest 82:511-512.

Gearhart, J. M. and R. B. Schlesinger. 1986. Sulfuric acid-induced airway hyper-responsiveness. Fundamental and Applied Toxicology 7:681-689.

Gerblich, A. A., H. Salik, and M. R. Schuyler. 1991. Dynamic T-cell changes in peripheral blood and bronchoalveolar lavage after antigen bronchoprovocation in asthmatics. Am Rev Respir Dis 143:533-537.

Gervais, P. 1966. L'asthme professionel dans l'industrie des matieres plastiques. Poumon et le Coeur 22:2199.

Goh, K., and R. V. M. Cestero. 1979. Chromosomal abnormalities in maintenance hemodialysis patients. J Med 10:16/7.

Goldsmith, J. R., and L. T. Friberg. 1977. Effects of air pollution on human health. Pp. 457-610 in Air Pollution. Vol. II The Effects of Air Pollution, A. C. Stern, ed. New York: Academic 59 Press, Inc.

Gregg, I. 1983. Epidemiological Aspects. Pp. 242-294 in Asthma, T. J. Clarke and S. Godfrey, eds., 2nd ed., London: Chapman and Hall.

Hakin, R. M., J. Breillatt, J. M. Lazarus, and F. K. Port. 1984. Complement activation and hypersensitivity reactions to dialysis membranes. New Eng J Med 311(14):878-882.

Hetzel, M. R., and T. J. H. Clark. 1983. Adult Asthma. Pp. 473-481 in Asthma, T. J. H. Clark and S. Godfred, eds. London: Chapman and Hall.
Hendrick, D. J., and D. J. Lane. 1975. Formalin asthma in hospital staff. Brit Med J 1:607-608. 60.

Hendrick, D. J., and D. J. Lane. 1977. Occupational formalin asthma. Brit J Ind Med 34:11-18.

Hoffmann, D. J., and E. L. Wynder. 1977. Organic particulate pollutants - chemical analysis and bioassays for carcinogenicity. Pp. 361-455 in Air Pollution. Vol. II The Effects of Air Pollution, A. C. Stern, ed. New York: Academic Press, Inc.

Hogg, J. C., A. L. James, and P. D. Pare. 1991. Evidence for inflammation in asthma. Am Rev Respir Dis 143:S39-S42.

Hoy, W. E., and R. V. M. Cestero. 1979. Eosinophilia in maintenance hemodialysis patients. J Dialysis 3(1):73-79.

Kensler, C. J., and S. P. Battista. 1963. Components of cigarette smoke with ciliary-depressant activity. New Eng J Med 269:1161-61 1166.
Kerfoot, E. J., and T. F. Mooney. 1975. Formaldehyde and paraformaldehyde study in funeral homes. Am Ind Hyg Assoc J 36:533-537.

Kilburn, K. H., R. Warshaw, C. T. Boylen, S. J. S. Johnson, B. Seidman, R. Sinclair, and T. Takaro. 1985. Pulmonary and neurobehavioural effects of formaldehyde exposure. Arch Environ Health 40:254-260.

Kotin, P. 1966. (need title) Can Cancer Conf 6:475. Kratochvil, I. 1971. Effects of formaldehyde on the health of workers in the crease-resistant clothing industry. Pracovni Lekarstvi 23(1):374-375.

Krzyzanowski, M., J. J. Quackenboss, and M. D. Lebowitz. 1990. Chronic respiratory effects of indoor formaldehyde exposure. Environ Res 52:117-125.

Kuz'menko, N. M., et al. 1975. Study of the sensitizing action of formaldehyde under the conditions of plastic manufacture. Vrachebnoe Del 6:131-134.

LaBelle, C. W., J. E. Long, and E. E. Christofano. 1955. Synergistic effects of aerosols. AMA Arch Ind Health 11:297-304.

Lebowitz, M. D., R. Barbee, and B. Burrows. 1984. Family concordance of IgE, atopy and disease. J Allergy Clin Immunol 73:259-266.

Loomis, T. A. 1979. Formaldehyde toxicity. Arch Pathol Lab Med 103:321-324.

Malorney, R., N. Reitbrock, and M. Schneider. 1965. Die oxydation des formaldehyds zu ameisensaure im blut, ein beitrag zum stoffwechsel des formaldehyds. Naunyn-Schmiedebergs Arch Pharmacol 250:419-436.

Marone, G. 1989. The role of mast cell and basophil activation in human allergic reactions. Eur Respir J 2(6 Suppl.):446S-455S.

Marsh, D. G., L. M. Lichtenstein, and P. S. Norman. 1972. Induction of IgE-mediated immediate hypersensitivity to group 1 rye grass pollen allergen and allergoids in non-allergic man. Immunology 22:1013.

McGue, M., J. W. Gerrand, M. D. Lebowitz, and D. C. Rao. 1989. Commingling in the distributions of immunologulin levels. Human Heredity 39(4):196-201.

Milby, T. H., M. M. Key, R. L. Gibson, et al. 1964. Chemical Hazards. Pp. 63-242 in Occupational Diseases, WM. M. Gafafer, ed. U.S. Department of Health, Education, and Welfare, Publication No. 1097, Washington,D.C., U.S. Government Printing Office.

Morin, N. R., P. E. Zeldin, Z. O. Kubinski, P. K. Bhattacharya, and H. Kubinski. 1977. Macromolecular complexes produced by chemical carcinogens and ultra-violet radiation. Cancer Res 37:3802-3814.

Moschandreas, D. J., R. H. Moyer, J. R. Ward, H. E. Rector, J. T. Schakenback, et al. 1980. An evaluation of formaldehyde problems in residental mobile homes. Pp. 1-146. Final Task Report I, GEOMET Report No.. ESF-797, April, for the Department of Housing and Urban Development. Washington, D.C.: EPA. 65

Nadel, J. A., and P. J. Barnes. 1984. Autonomic regulation of the airways. Am Rev Med 35:451-467. Nagai, S., H. Aung, M. Takeuchi, K. Kusume, and T. Izumi. 1991. IL-1 and IL-1 inhibitory activity in the culture supernatants of alveolar macrophages from patients with interstitial lung diseases. Chest 99:674-680.

Nagornyi, P. A., Z. A. Sudakova, and S. M. Schablenko. 1979. General toxic and allergenic action of formaldehyde. Gigiena Truda i professional'nye Zabolevaniya 1:27-30. National Research Council (NRC). 1981a. Formaldehyde and other aldehydes. National Academy Press, Washington, D.C.

National Research Council (NRC). 1981b. Indoor Pollutions. National Academy Press, Washington, D.C. Nemery, B. 1990. Metal toxicity and the respiratory tract. Eur Respir J 3:202-219. Newman-Taylor, A. J., and R. J. Davies. 1981. Inhalation Challenge Testing. Pp. 143-168, in Occupational Lung Diseses, H. Weill and M. Turner-Warwick, eds. New York: Marcel Dekker. Newman-Taylor, A., and R. D. Tee. 1990. Environmental and occupational asthma. Exposure assessment. Chest 98(5 Suppl.):209S-215S.

Nordman, H., H. Keskinen, and M. Tuppurainen. 1985. Formaldehyde asthma - rare or overlooked? J Allergy Clin Immunol 75:81-99. Nova, M. M. H., and R. G. Touraine. 1956. Asthme au formol, Pp. 293-294 in Societe De Medecine Du Travail, Lyon.

Odom, R. G., and H. I. Maiback. 1977. 1977. Contact urticaria: a different contact dermatitis. Pp. 441-452 in Advances in Modern Toxicology, Vol. 4, Dermatotoxicology and Pharmacology, F. N. Marzulli and H. I. Maibach, eds. Washington, DC: Hemisphere Publishing.

Omini, C., G. Brunelli, A. Hernandez, and L. Daffonchio. 1989. Role of the mediators in pulmonary hyperreactivity: the cocktail interaction hypothesis. Eur Respir J 2(6 Suppl.):493S-496S. Orringer, E. P., and W. D. Mattern. 1976. Formaldehyde-induced hemolysis during chronic hemodialysis. New Eng J Med 294(26):1416-1420.

Ostapovich, I. K. 1975. Conditions of respiratory route exposure to sulfur dioxide and formaldehyde and subsequent sensitization. Gigyena I Sanitariya 2:9-13.

Otsuka, H., J. A. Denburg, and J. Dolovich. 1985. Heterogeneity of metachromatic cells in human nose: significance of mucosal mast cells. J Allergy Clin Immunol 76:695-702.

Otsuka, K., J. Dolovich, A. D. Befus, et al. 1986. Basophilic cell progenitors, nasal metachromatic cells and peripheral blood basophils in ragweed allergic rhinitis. J Allergy Clin Immunol 78:365-371.

Otsuka, H., J. Dolovich, M. Richardson, J. Bienenstock, and J. A. Denburg. 1987. Metachromatic cell progenitors and specific growth and differentiation factors in human nasal mucosa and polyps. Am Rev Respir Dis 136:710-717.

Paggiaro, P. L., E. Bacci, N. Pulera, P. Bernard, F. L. Dente, D. Talini, and C. Giuntini. 1989. Vagal reflexes and asthma. Eur Respir J 2(6 Suppl.):502S-507S.

Patterson, R., L. C. Grammer, C. R. Zeiss, K. E. Harris, and M. A. Shaughnessy. 1990. Use of immunologic technology in the diagnosis of environmental and occupational immunologic lung disease. Chest 98(5 Suppl.):206S-208S.

Paliard, F., L. Roche, C. Exbrayat, and E. Sprunk. 1949. Chronic asthma due to formaldehyde. Arch Mal Prof 10:528.

Pirila, V., and O.Kilpio. 1949. On dermatitis caused by formaldehyde 70 and its compounds. Am Med Intern Fenn 38:38-51.

Popa, V, D. Teculescu, D. Stanescu, and N. Gavrilescu. 1969. Bronchial asthma and asthmatic bronchitis determined by simple chemicals. Dis Chest 56(5):395-404.

Postma, D. S., J. J. Keyzer, G. H. Koeter, H. J. Sluiter, and K. DeVries. 1985. Influence of the parasympathetic and sympathetic nervous system on nocturnal bronchial obstruction. Clin Sci 69:251-258.

Quackenboss, J. J., M. D. Lebowitz, C. Hayes, and C. L. Young. 1989. Respiratory responses to indoor/outdoor air pollutants: combustion products, formaldehyde, and particulate matter. Pp. 280-293 in Combustion Processes and the Quality of the Indoor Air Environment, J. Harper, ed. Pittsburgh: Air Pollution Control Association.

Quackenboss, J. J., M. D. Lebowitz, and C. Hayes. 1989a. Epidemiological study of respiratory response to indoor/outdoor air quality. Environ Int 15:493-502.

Ramsdale, E. H., M. M. Morris, R. S. Roberts, and F. E. Hargreave. 1985. Asymptomatic bronchial hyperresponsiveness in rhinitis. J. Allergy Clin. Immunol. 95(5):573-577.

Randem, B., M. H. Smolensky, B. Hsi, D. Albright, and S. Burge. 1987. Field survey of circadian rhythm in PEF of electronics workers suffering from colophony-induced asthma. Chronobiology International 4(2):263-271.

Reed, C. E. 1988. Basic mechanisms of asthma. Role of inflammation. Chest 94(1):175-177 Rockel, A., B. Klinke , J. Hertel, X. Baur, C. Thiel, S. Abdelhamid, P. Fiegel, and D. Walb. 1989. Allergy to dialysis materials. Nephrol Dial Transplant 4:646-652.

Rose, N. R., F. Milgram, and C. J. van Oss. 1973. Principles of Immunology. New York: MacMillan Publishin Co.,Inc.

Rossi, G. A., O. Sacco, F. Vassallo, S. Lantero, A. Morelli, U. Benatti, and G. Damiani. 1989. Oxidative metabolism of human peripheral blood eosinophils and neutrophils: H2O production after stimulation with phorbol myristate acetate and immune complexes. Eur Respir J 2(6 Suppl.):435S-440S.

Roth, W. G. 1969. Tylotic palmer and planter eczema caused by steam ironing clothes containing formaldehyde. Berufs-Dermatosen 17:263-267.

Saetta, M., L. M. Fabbri, D. Danieli, G. Picotti, and L. Allegra. 1989. Pathology of bronchial asthma and animal models of asthma. Eur Respir J 2(6 Suppl.):477S-482S.

Sakula, A. 1975. Formalin asthma in hospital laboratory staff. Lancet 2:816.

Sandler, S. G., R. Sharon, M. Bush, M. Stroup, and B. Sabo. 1979. Formaldehyde-related antibodies in hemodialysis patients. Transfusion 19(6):682-687.

Semenzato, G. 1991. Immunology of interstitial lung diseases: cellular events taking place in the lung of sarcoidosis, hypersensitivity pneumonitis and HIV infection. Eur Respir J 4:94-102.

Semin, Y.A., E. N. Kolomyitseva, and A. M. Poverennyi. 1974. Effects of products of the reaction of formaldehyde and amino acids on nucleotides and DNA. Molekulyarnaya Biologiya 8:276.

Schlueter, D.P. 1982. Infiltrative lung disease hypersensitivity pneumonitis. J Allergy Clin Immunol 70(1):50-55.

Schoenberg, J. B., and C. A. Mitchell. 1975. Airway disease caused by phenolic (Phenol-formaldehyde) resin exposure. Arch Environ Health 30:574-577.

Shim, C., M. H. Williams, Jr. 1986. Effect of odors in asthma. Amer J Med 80:18-22.

Shipkovitz, H. D. 1968. Formaldehyde vapor emissions in the permanent-press fabrics industry. Report No. TR-52. U.S. Public Health Service, Cincinnati, Ohio.

Skerfving, S., B. Akesson, and B. G. Simonsson. 1980. "Meatwrappers' asthma" caused by thermal degradation products of polythylene. Lancet 1:211.

Sneddon, I. B. 1968. Dermatitis in an intermittent haemodialysis unit. Brit Med J 1:183-184.

Spassovski, M. 1976. Health hazards in the production and processing of some fibers, resins, and plastics in Bulgaria. Environ Health Perspec 17:199-202.

Stokinger, H. E., and D. L. Coffin. 1968. A formaldehyde and its homologs, Pp. 483-484 in Volume I Air Pollution, A. C. Stern, ed. New York: Academic Press.

Tara, M. S. 1966. Intolerance? Sensitization? Formaldehyde-induced pulmonary allergy? Arch Mal 17:66-68.

Tarlo, S. M., and I. Broder. 1989. Irritant-induced occupational asthma. Chest 96(2):297-300.

Tathirov, M. T. 1974. Experimental study of the combined effect of six atmospheric pollutants on the human body. Gig Sanit 5:100-102.

Thrasher, J. D., R. Madison, A. Broughthon, and Z. Gard. 1989. Building-related illness and antibodies to albumin conjugates of formaldehyde, toluene diisocyanate and trimellitic anhydride. Am J Ind Med 15:187-195.

Thrasher, J. D., A. Broughton, and R. Madison. 1990. Immune activation and autoantibodies in humans with long-term inhalation exposure to formaldehyde. Archives Environ Hlth 45(4):217-223.

Thurmond, L. M., and J. H. Dean. 1988. Immunological responses following inhalation exposure to chemical hazards. Pp. 375-392 in Toxicology of the Lung, D. E. Gardner, J. D. Crapo, and E. J. Massaro, eds. New York: Raven Press.

Trinkler, H. 1968. Working with formaldehyde. Medizinische Laboratorium (Stuttgart) 21:283. Turner-Warwick, M., ed. 1978. Immunology of the Lung. London: E. Arnold (Publishers) Ltd.

Tuthill, R. W. 1984. Woodstoves, formaldehyde, and respiratory disease. Amer J Epidemiol 120(6):952-955.

Vaughn, W. T. 1939. The Practice of Allergy. P. 677. St. Louis: C. V. Mosby. Walker, J. F. 1964. Formaldehyde. Pp. 484. New York: Rinehold Publishing Corp.

Weber, A. 1984. Acute effects of environmental tobacco smoke. Eur J Resp Dis 68(Suppl.133):98-108.

Wegman, D. H., and E. A. Eisen. 1990. Acute irritants. More than a nuisance. Chest 97(4):773-775.

Weill, H., and M. Turner-Warwick, eds. 1981. Occupational Lung Diseases: Research Approaches and Methods. New York:Marcel Dekker, Inc.

WHO/EURO. 1986. Indoor Air Quality: Radon and Formaldehyde. World Health Organization Regional Office for Europe, Environ Health

WHO/EURO. 1987. Air Quality Guidelines. World Health Organization Regional Office for Europe, Vol. 2, Chap. 5, Copenhagen/Geneva.

WHO/EURO. 1991. Indoor Air Quality: Combustion Products, Copenhagen, in press. Yefremov, G. G. 1970. The state of the upper respiratory tract in formaldehyde production employees. Zh Ushn Nos Gorl Bolezn 30:11-15.

Zaeva, G. N., et al. 1968. Materials for the revision of the maximum 80 permissible concentration of formaldehyde in the interior atmosphere of industrial premises. Gigiena Truda Professional'nye Zab 12(7):16-20.

Multiple Chemical Sensitivity-What is It?

Roy L. DeHart

"If you don't know where you are going any road will get you there."

–Anonymous

This workshop on multiple chemical sensitivity accomplishes many objectives; perhaps most significant was gathering many professionals with widely divergent opinions regarding MCS together. One important objective unfortunately must await another time: the definition of MCS.

The inability to reach concensus of such an important aspect of this phenomenon speaks volumes on the controversy that is MCS. This controversy is neither new nor unstated. Several disciplines in organized medicine have addressed the issues and contributed to the controversy. Four of these positions are presented for review.

The California Medical Association established a Task Force on Clinical Ecology under its Scientific Board. The report of the Task Force was published in February, 1986 which in part states:

The task force collected material as for any subject review and included all information supplied by individual clinical ecologists and by their professional organizations. There was extensive description of the basic hypotheses of clinical ecology, and an ample and varied collection of anecdotal reports and individual patient testimonials. In contrast, there was a surprising paucity of published studies to prove or disprove clinical ecology hypotheses. Critical analyses of patients and cohorts, detailed data collection, validation and confirming laboratory assays were not provided.

No convincing evidence was found that patients treated by clinical ecologists have unique, recognizable syndromes, that the diagnostic tests employed are efficacious and reliable or that the treatments used are effective. Even though clinical ecology has existed for approximately 50 years, only a few studies have been conducted that are scientifically sound. Most have such serious methodological flaws as to make their conclusions unacceptable. Those few studies that used scientifically sound methods have provided evidence that the effectiveness of certain treatment methods used by clinical ecologists is based principally on placebo response.

Undoubtedly, some patients suffer from illnesses that cannot be readily diagnosed and

35

for which only supportive treatments exist. It may even be true that some or all of the hypotheses and treatments proposed by clinical ecologists are valid but we found no evidence to support them. These hypotheses and treatments should be subjected to modern, scientific methods of evaluation. We think that this can be done provided genuine interest exists.

The task force is concerned that unproved diagnostic tests are being widely used by clinical ecologists in what may be incorrect or inappropriate applications. Decisions made on the basis of these tests can lead to misdiagnosis, resulting in patients being denied other supportive treatments and becoming psychologically dependent, believing themselves seriously and chronically impaired. This possibility underscores the need for more adequate scientific studies to prove or disprove the value of clinical ecology tests and treatments. To consider the current practice of clinical ecology experimental is misleading, however. It can only be considered experimental when its practitioners adhere to scientifically sound research protocols and inform their patients about the investigative nature of their practice.[1]

The same year the Executive Committee of the American Academy of Allergy and Immunology issued a position statement on Clinical Ecology. The critique follows:

The environment is very important in the lives of every human being. Environmental factors, such as chemicals and pollutants, have been demonstrated to influence health. The idea that the environment is responsible for a multitude of human health problems is most appealing. However, to present such ideas as facts, conclusions, or even likely mechanisms without adequate support, is poor medical practice.

The theoretical basis for ecologic illness in the present context has not been established as factual, nor is there satisfactory evidence to support the actual existence of "immune system disregulation" or maladaptation. There is no clear evidence that many of the symptoms noted above are related to allergy, sensitivity, toxicity, or any other type of reaction from foods, water, chemicals, pollutants, viruses, and bacteria in the context presented. Properly controlled studies defining objective parameters of illness, properly controlled evaluation of the treatment modalities, and appropriate patient assessment have not been done. Anecdotal articles do not constitute sufficient evidence of a cause-and-effect relationship between symptoms and environmental exposure. The major techniques used by the clinical ecologists are controversial and unproven. The American Academy of Allergy and Immunology has previously published position statements concerning subcutaneous and sublingual provocation neutralization procedures and found them to be unproven. More recent review of new data submitted by a number of clinical ecologists to the Practice Standards Committee of the Academy has not changed that recommendation. There are no adequate studies of the cyclic diets, elimination diets, injection therapy with chemicals, or even the environmentally controlled units to substantiate their use. Many of the patients are reported to have a normal physical examination and normal laboratory tests.

There are no immunologic data to support the dogma of the clinical ecologists. To suggest that these patients lack suppressor T cell function has not been supported by published data. The suggestion that neutralization therapy can provide rapid relief within minutes or hours cannot be supported by controlled clinical studies or immunologic data.[2]

The American College of Physicians published a position paper on Clinical Ecology in 1989. The summary from that publication is quoted:

Clinical ecologists propose the existence of a unique illness in which multiple environmental chemicals, foods, drugs, and endogenous *C. albicans* have a toxic effect on the immune system, thereby adversely affecting other bodily functions. The proposal uses some concepts that superficially resemble those that apply to clinical allergy and toxicology and others that are novel.

Review of the clinical ecology literature provides inadequate support for the beliefs and practices of clinical ecology. The existence of an environmental illness as presented in clinical ecology theory must be questioned because of the lack of a clinical definition. Diagnoses and treatments involve procedures of no proven efficacy.

Case reports by clinical ecologists and evaluation of these patients by other physicians indicate that this diagnosis is applied most frequently to persons with symptoms of physiologic (somatic) or psychologic dysfunction, or both. Proof of cause-effect relations between environmental factors and symptoms of "environmental illness" is particularly difficult because clinical ecologists implicate such a broad range of agents, including chemicals, foods, hormones, and microorganisms. Most patients are believed to react to multiple environmental substances by any route of exposure, and some are said to be intolerant to the entire environment, the so-called "total allergy syndrome."

The principal method of proof cited by clinical ecologists for the existence of "environmental illness" is the symptom-provocation test used in diagnosis of individual cases after the condition is suspected because of a history of symptoms and suspected causes. Published studies on the provocation test employed widely different subject-selection methods and outcome-measurement criteria. All were seriously flawed by the absence of matched patient-control groups, absence or inadequacy of placebo, and failure to achieve or document randomness of trials. Not surprisingly, therefore, the conclusions from these studies are conflicting.

Those studies reporting results of immunologic tests are insufficient to address theories of environmental illness; the number of cases is small and selection criteria are not clear. Enumeration of lymphocyte subsets and quantitation of serum immunoglobulin and complement levels in patients with "environmental illness" have not yielded clear-cut evidence of immunologic abnormality.

Clinical ecologists use a treatment program that includes avoidance of environmental chemicals, rotation of foods in the diet, and neutralization of symptoms with injected or sublingual extracts. Except for small-dose oral nystatin, which is used for treatment of patients with the candida hypersensitivity syndrome, drug therapy is intentionally avoided, although some clinical ecologists recommend mineral salts, oxygen, vitamins, minerals, and antioxidants for relief of symptoms. There are only two controlled studies on neutralization therapy. One evaluated only eight patients for 20 days; the other was a study of dust allergy in patients with allergic rhinitis, not "environmental illness." To date, there have been no controlled studies on diet therapy, environmental elimination therapy, or nystatin in patients with candida hypersensitivity. Thus there is no body of evidence that clinical ecology treatment measures, singly or in combination, are effective.

Others who have reviewed the diagnostic and therapeutic methods of clinical ecology have also arrived at the conclusion that these methods are of unproven value. Potential adverse effects from these procedures and the costs of clinical ecology diagnosis and treatment have not been evaluated.[3] More recently, on April 28, 1991, the American College of Occupational Medicine issued the following position statement on multiple chemical hypersensitivity syndrome:

Multiple Chemical Hypersensitivity Syndrome is one of over 20 names or descriptions (20th century disease, Environmental Illness, Total Allergy Syndrome, Chemical AIDS, etc.) given to a clinical complex characterized by recurrent polysystem symptoms often without observable physical findings or laboratory abnormalities believed to be associated with repeated exposure to specific biological, chemical or physical agents. Practitioners involved in the diagnosis and treatment of this phenomenon may identify themselves as clinical ecologists or environmental medicine specialists.

The pathophysiological mechanism described by these practitioners do not, in general, conform to what is currently known of human biological functions. To explain the phenomenon, these practitioners draw on new or modified mechanisms such as "total body load, spreading, and switching." The scientific foundation for managing patients with this syndrome has yet to be established by traditional clinical investigative activities that withstand critical peer review.

The following medical associations have carefully reviewed the available scientific evidence regarding this phenomenon:

The American Academy of Allergy and Immunology

The California Medical Association

The American College of Physicians

Their conclusions have been similar and state that the scientific and clinical evidence supporting the pathophysiological mechanisms and treatment regimens as articulated by these practitioners is lacking. It is the position of the American College of Occupational Medicine that the Multiple Chemical Hypersensitivity Syndrome is presently an unproven hypothesis and current treatment methods represent an experimental methodology. The College supports scientific research into the phenomenon to help explain and better describe its pathophysiological features and define appropriate clinical interventions. This research should adhere to established principles of scientific inquiry and the results submitted for publication in recognized peer-reviewed journals.[4]

In each of these position statements there is one integrating theme–the clear need for rigorous scientific placebo-controlled, double-blind investigations that are subject to exacting peer review. Although the working group was unable to characterize the patient with multiple chemical sensitivity, the investigators and scientists must provide a definition as part of protocol development. Case definition and criteria for a diagnosis must be explicit if the study population is to be appropriately identified. Further, if one is to define the mechanism of MCS or to conduct epidemiological research focused on MCS, case definition is not an option, but a necessity.

If the question cannot be answered as to what MCS is, how can there be approval of research protocols or acceptance of investigative results? In order to appropriately address the controversies surrounding this phenomenon we must know where we're going!

REFERENCES

California Medical Association. *Clinical Ecology–A Critical Appraisal.* Western Journal of Medicine; 144:2,239-49, 1986.

American Academy of Allergy and Immunology. *Clinical Ecology.* Journal of Allergy and Clinical Immunology 78:2,269-71; 1986.

American College of Physicians. *Clinical Ecology.* Annals of Internal Medicine 111:2,168-78; 1989.

American College of Occupational Medicine. *Multiple Chemical Hypersensitivity Syndrome.* Board minutes, April 28, 1991.

Case Definitions for
Multiple Chemical Sensitivity[1]

Nicholas A. Ashford and Claudia S. Miller

TERMINOLOGY

A wide array of names has been applied to the syndromes suffered by patients with heightened reactivity to chemicals (Table 1). Each name has specific implications regarding the underlying cause, mechanism, or manifestations of the disease, and they overlap. A major hindrance in achieving scientific respectability has been the difficulty in agreeing upon a definition for this condition (or conditions). It should be realized that no single case definition, even if agreed upon in one context--e.g., for diagnosis purposes--will suffice for use in other contexts. It is important to distinguish case definitions for use in diagnosis, epidemiological studies, research on the nature of the condition, regulatory standard-setting, compensation awards, and situations requiring alternative employment or housing. The social and political consequences of the use of a particular definition in a specific context requires the case definition to be carefully constructed.

Cullen (1987b) has emphasized the importance of establishing a uniform case definition before meaningful epidemiologic studies can be undertaken, but cautions, "However constructed, the goal of descriptive studies must be refinement of the diagnostic criteria, in particular the very tentative boundaries with other diagnostic entities such as allergic, anxiety, panic and post-traumatic stress disorders, and physiologic sequelae of central nervous system (CNS) intoxication or injury, especially by organic solvents." He acknowledges possible overlap among these entities and offers the following case definition:

Multiple chemical sensitivities (MCS) is an acquired disorder characterized by recurrent symptoms, referable to multiple organ systems, occurring in response to demonstrable exposure to many chemically unrelated compounds at doses far below those established in the general population to cause harmful effects. No single widely accepted test of physiologic function can be shown to correlate with symptoms. (Cullen 1987a)

This case definition, intended for epidemiological use, is intentionally narrow. Cullen

---

[1] Excerpted in part from N.A. Ashford and C.S. Miller, Chemical Exposures: Low Levels and High Stakes, Van Nostrand Reinhold, New York 1991.

TABLE 1

Cause	Mechanism	Effect
Environmentally induced illness	Immunologic illness	Multiple chemical sensitivities (MCS)
	Immunotoxicity	
Chemically induced (or acquired) hypersusceptibility	Immune dysfunction	Multiple chemical sensitivity syndrome
Chemically acquired immune deficiency syndrome (chemical AIDS)	Immune dysregulation	Chemical hypersensitivity syndrome
	Conditioned odor response	
		Universal allergy
The petro-chemical problem	Fear/anxiety	
		20th-century illness
	Mass psychogenic illness	
		Total allergy syndrome
	Various psychiatric disorders	
		Environmental allergy or illness
		Cerebral allergy
		Environmental maladaptation syndrome
		Food and chemical sensitivity

This case definition, intended for epidemiological use, is intentionally narrow. Cullen excludes persons who react to substances no one else in aware of on the basis that such individuals may be delusional and excludes persons who have bronchospasm, vasospasm, seizures, or "any other reversible lesion" that can be identified and specifically treated. Clinical ecologists, however, would argue that persons with bronchospasm, vasospasm, seizures, and other illnesses excluded by Cullen may well have the chemical sensitivity problem. Each issue of the clinical ecologists' journal, Clinical Ecology, contains the following definition:

Ecologic illness is a chronic multi-system disorder, usually polysymptomatic, caused by adverse reactions to environmental incitants, modified by individual susceptibility and specific adaptation. The incitants are present in air, water, food, drugs and our habitat.

Although the patients the clinical ecologists and Cullen see are demographically divergent, the definitions of their illnesses are remarkably alike. Both describe the chemically sensitive patient in similar terms. [See Miller and Ashford, "Allergy and Multiple Chemical Sensitivity (MCS) Distinguished" in this report for a discussion of sensitive populations.]

However, what is sorely needed is an objective test that can be applied in each individual case to determine, incontrovertibly, whether a particular person has multiple chemical sensitivities.

Given the multitude of environmental exposures (both chemical and food) that allegedly can result in a seemingly endless array of physical and mental syndromes and the frequent absence of findings on routine physical examination, the practitioner who sees these patients with their divergent and unfamiliar litany of complaints is at great disadvantage in trying to diagnose the condition.

To circumvent this problem, we propose the following operational definition of multiple chemical sensitivity for diagnostic purposes, a definition that is based upon environmental testing:

The patient with multiple chemical sensitivities can be discovered by removal from the suspected offending agents and by rechallenge, after an appropriate interval, under strictly controlled environmental conditions. Causality is inferred by the clearing of symptoms with removal from the offending environment and recurrence of symptoms with specific challenge.

Challenges conducted for research purposes should be performed in a double-blind, placebo-controlled manner.

This operational definition is essential to resolving, once and for all, the debate about whether an individual's symptoms are or are not environmentally induced. An environmental unit is necessary for scientific validation of the concept of chemical sensitivity. Because of the expense and time required by patients and physicians alike, we are not arguing that the unit be used for all patients. Such stringent measures are not necessary for most patients. For severe cases, however, no alternative is available at present, and only from firsthand observation of hospitalized patients can physicians have the opportunity to understand this illness better. In time, as more clinical data on these patients accumulate, physicians may be able to diagnose this disorder on the basis of the patient's history and a few key laboratory tests. For now, reliance must be placed on rigorous study in an environmental unit. Ultimately a phenomenological definition may emerge that allows physicians to diagnose, at least tentatively, chemical sensitivity based on a history of a specific sensitizing event (such as a pesticide exposure) followed by evidence of chemical and food sensitivities, multisystem effects, improvement after avoidance of exposure, and similar experiences of persons with like histories.

Criteria for the selection of cases *for research purposes* are:

- Sensitivity to chemicals, i.e., symptoms or signs related to chemical exposures at levels tolerated by the population at large that is distinct from such well recognized hypersensitivity phenomena as IgE-mediated immediate hypersensitivity reactions, contact dermatitis, and hypersensitivity pneumonitis.
- Sensitivity, expressed as symptoms and signs, in one or more organ systems.
- Symptoms and signs that wax and wane with exposures.

It is not necessary to identify a chemical exposure associated with the onset of the condition. Preexistent or concurrent conditions, e.g. asthma, arthritis, somatization disorder or depression, should not exclude patients from consideration.

The selection of subjects for research protocols should depend on the specific hypotheses to be tested. Identifiable populations should include but not be restricted to:

- Symptom or sign based: Patients with reactivity to environmental chemicals, either through self-reporting or meeting case selection criteria.
- Disease based: Patients with specific diseases that are suspected to be caused by or exacerbated by chemical exposures.
- Exposure based: Groups characterized by a common exposure, such as workers at a specific factory, occupants of a particular building, or residents of a contaminated community.
- Population based: Groups such as school children or a random community sample.

Appropriate comparison groups should be chosen in each case.

Whatever case definitions emerge for diagnostic or research purposes, decisions concerning "safe levels" of chemical exposure for regulatory standard-setting, criteria of eligibility for disability awards under workers' compensation or social security, and criteria of eligibility for alternative employment or housing ultimately need to be based on suitable definitions. These definitions should be expected to differ in exactness and stringency, and may be in flux until more research results are obtained.

REFERENCES

Cullen, M., "The Worker with Multiple Chemical Sensitivities: An Overview." In: Cullen, M. (ed.) Workers with Multiple Chemical Sensitivities, Occupational Medicine: State of the Art Reviews (1987a), Hanley and Belfus, Philadelphia, 2(4):655-662.

Cullen.M., "Multiple Chemical Sensitivities: Summary and Directions for Future Investigators." In: Cullen, M. (ed.) Workers with Multiple Chemical Sensitivities, Occupational Medicine: State of the Art Reviews (1987b), Hanley and Belfus, Philadelphia, 2(4):801-804.

Allergy and Multiple Chemical Sensitivities Distinguished[1]

Claudia S. Miller and Nicholas A. Ashford

When von Pirquet coined the term "allergy" in 1906, he defined it as "altered reactivity". Thus the word "allergy" as originally conceived encompassed both immunity and hypersensitivity. In 1925 European allergists influenced their American colleagues to redefine "allergy" in the context of antibodies and antigens, effectively excluding hypersensitivity on any other basis.

Forty years ago, Theron Randolph, a classically trained allergist, noted that a cosmetic saleswoman he had been seeing for rhinitis, asthma, headache, fatigue, irritability, depression, marked weight fluctuation and intermittent episodic loss of consciousness developed her symptoms following exposure to gas, oil, coal and their combustion products (Randolph, 1987, pp. 73-76). Randolph developed a diagnostic-therapeutic maneuver which consisted of removing the patient from all suspected environmental exposures and subsequently reintroducing single elements of the environment, one at a time, while observing for changes in the patient's condition (Randolph, 1960). Although what he observed in his patients appeared to be some kind of hypersensitivity, it was not allergy. Subsequently, Randolph and other physicians known as clinical ecologists published clinical descriptions of patients with polysomatic complaints, frequently including mood and cognition difficulties, triggered by a wide variety of chemical exposures, but especially petrochemical exposures, and often with concomitant food and drug intolerances. These clinical descriptions bear striking resemblance to today's MCS patients, many of whom have never heard of clinical ecology.

A review of the literature on exposure to low levels of chemicals reveals four groups or clusters of people among whom individuals with heightened reactivity have been reported (Ashford and Miller, 1989):

1. Industrial workers
2. Occupants of "tight buildings," including office workers and school-children

---

[1]Excerpted from Ashford, N. A. and Miller, C. S. 1991. Chemical Exposures: Low Levels and High Stakes. New York: Van Nostrand Reinhold.

3.   Residents of communities whose air or water has been contaminated by chemicals
4.   Individuals who have had personal and unique exposures to various chemicals in domestic indoor air, pesticides, drugs, and consumer products

These four groups are listed for comparison in Table 1. Note that they differ in professional and educational attainment, age and sex, and the mix and levels of chemicals to which they are exposed, but that all have multiple symptoms involving multiple organ systems with marked variability in the type and degree of those symptoms. Symptoms are often "subjective". For example, central nervous system (CNS) symptoms such as difficulty concentrating or irritability are common, and physical examinations are frequently unremarkable for individuals in each category.

TABLE 1

Chemically Sensitive Groups

Group	Nature of Exposure	Demographics
Industrial workers	Acute and chronic exposure to industrial chemicals	Primarily males: blue collar; 20 to 65 years old
Tight-building occupants	Off-gassing from construction materials, office equipment or supplies; tobacco smoke; inadequate ventilation	Females more than males; white collar office workers and professionals; 20 to 65 years old; schoolchildren
Contaminated communities	Toxic waste sites; aerial pesticide spraying; ground water conatmination; air contamination by nearby industry and other community exposures	All ages, male and female; children or infants may be affected first or most; pregnant women with possible effects on fetuses; middle to lower class
Individuals	Heterogeneous; indoor air (domestic), consumer products, drugs, and pesticides	70-80% females; 30 to 50 years old (Johnson and Rea 1989); white middle to upper middle class and professionals

Many affected individuals report a major precipitating (inducing or "sensitizing") exposure which marked the onset of their chemical sensitivities. In one survey of 6,800 persons claiming to be chemically sensitive, 80 percent asserted that they knew "when, where, with what, and how they were made ill" (National Foundation for the Chemically Hypersensitive, 1989). Of the 80 percent, 60 percent (that is, almost half of those who replied) blamed pesticides. The respondents to the survey were self selected, and the result must be interpreted with caution. Nevertheless, the results suggest that future surveys of persons with different exposure histories and symptoms might contribute to an understanding of underlying mechanisms and causes.

In some chemically sensitive patients no single, identifiable, "high-level" exposure seems to have been associated with the onset of their difficulties. Exposures could have occurred but were not recognized or remembered. Some observers suggest that repetitive or cumulative lower level exposure events may lead to the development of sensitivities. Still others implicate genetic predisposition, pregnancy, major surgery with anesthesia, physical trauma, or major psychological stress as contributors to the illness. Based upon the increasing number of outbreaks of sick building syndrome, increased reporting of symptoms in contaminated communities to state health departments, increased recognition of problems in the industrial workplace, and the increasing numbers of physicians treating chemically related sensitivities, the existing evidence does suggest that chemical sensitivity is on the rise and could become a large problem with significant economic consequences related to the disablement of productive members of society.

The fact that the demographically diverse groups listed in Table 1 share similar patterns of illness (that is, onset after a major chemical exposure; subsequent hyperreactivity to low levels of a variety of chemicals commonly encountered in the environment such as cigarette smoke, perfume, and traffic exhaust; and multisystem complaints with frequent mood, memory and concentration difficulties) suggests that the problem may be real.

In addition, the temporal cohesiveness of onset of illness within groups of individuals sharing a recognized, major chemical exposure event, for example, the development of symptoms in several family members, co-workers or community members exposed simultaneously, help point to the problem as potentially real in those circumstances.

Although a definitive and accurate picture is yet to come, at this time these pieces - viewed collectively - provide sufficient evidence to conclude that chemical sensitivity does exist as a serious health and environmental problem.

The different meanings of the term "sensitivity" are at least in part responsible for the confusion surrounding chemical sensitivity. In the classical, toxicological use of the word "sensitivity", those individuals who require relatively lower doses to induce a particular response are said to be more sensitive than those who would require relatively higher doses before experiencing the same response. A hypothetical distribution of sensitivities, that is, the minimum doses necessary to cause individuals in a population to exhibit a harmful effect, is shown in curve A in Figure 1-1. (If we plot the cumulative number of individuals who exhibit a particular response as a function of dose, we generate a population dose-response curve; see curve A in Figure 1-2). This distribution describes the traditional toxicological concept of sensitivity. Curve A in Figure 1-1 illustrates that health effects of classical diseases are seen in a significant portion of the normal population at a certain dose; the sensitive and resilient populations are found in the tails of the distribution.

A second meaning of the word "sensitivity" appears in the context of classical IgE-mediated allergy (atopy). IgE is one of five classes of antibodies made by the body, and is, from the perspective of classically allergic individuals, the most important antibody. Atopic individuals have IgE directed against specific environmental incitants, such as ragweed or bee venom. Positive skin tests in these individuals correlate with a rapid onset of symptoms when they are actually exposed to those allergens. The atopic individual exhibits a reaction whereas non-allergic individuals do not, even at the highest doses normally found in the environment. A hypothetical sensitivity distribution for an atopic effect is shown in curve B of Figure 1-1, and the dose-response curve derived from that distribution is found in curve B

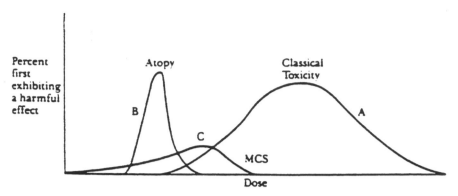

FIGURE 1-1   Hypothetical distribution of different types of sensitivities as a function of dose. Curve A is a sensitivity distribution for classical toxicity, e.g., to lead or a solvent. Sensitive individuals are found in the left-hand tail of the distribution. Curve B is a sensitivity distribution of atopic or allergic individuals in the population who are sensitive to an allergen, e.g., ragweed or bee venom. Curve C is a sensitivity distribution for individuals with multiple chemical sensitivities who, because they are already sensitized, subsequently respond to particular incitants, e.g., formaldehyde or phenol.

of Figure 1-2. Allergists include in the term "allergy" well-characterized immune responses that result from industrial exposure to certain chemicals, such as nickel or toluene diisocyanate (TDI). Most allergists refer to such responses as "chemical sensitivity", and reserve this term for responses that have a distinct immunological basis, preferring to use a term such as "chemical intolerance" for nonimmunological responses to chemicals.

Patients suffering from multiple chemical sensitivities (MCS) may be exhibiting a third and entirely different type of sensitivity. Their health problems often, but not always, appear to originate with some acute or traumatic exposure, after which the triggering of symptoms and observed sensitivities occur at very low levels of chemical exposure. The inducing chemical or substance may or may not be the same as the substances that thereafter provoke or "trigger" responses. (Sometimes the inducing substance is described as "sensitizing" and the individual affected as a "sensitized" person). Reactions may sometimes be observed at incitant levels similar to those to which classically sensitive and atopic patients respond. Unlike classical toxicity, however, here the effects of low-level exposures are not simply those effects observed in normal populations at higher doses. The fact that normal persons - for example, most doctors - do not experience even at higher levels of exposure those symptoms that chemically sensitive patients describe at much lower levels of exposure probably helps to explain the reluctance of some physicians to believe that the problems are physical in nature. (Although this also describes atopy, in this case the sensitivity is not IgE mediated). To compound the problem of physician acceptance of this illness, multiple organ systems may be affected, and multiple substances may trigger the effects. Over time, sensitivities seem to spread, in terms of both the types of triggering substances and the systems affected (Randolph, 1962, pp. 98 and 119). Avoidance of the offending substances is usually effective but much more difficult to achieve for these patients than for classically sensitive patients because symptoms may occur at extremely low doses

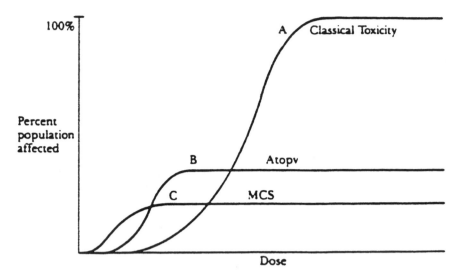

FIGURE 1-2   Hypothetical population dose-response curves for different effects. Curve A is a
cumulative dose-response cur e for classical toxicity, e.g., to lead or a solvent. Curve B is a
cumulative dose-response curve for atopic or allergic individuals in the population who are sensitive
to an allergen, e.g., ragweed or bee venom. Curve C is a cumulative dose-response curve for
individuals with multiple chemical sensitivities who, because they are already sensitized,
subsequently respond to particular incitants, e.g., formaldehyde or phenol.

and the exposures are manifold and ubiquitous. Adaptation to chronic low-level exposure
with consequent "masking" of symptoms (discussed more fully later) may make it
exceedingly difficult to discover these sensitivities and unravel the multifactorial triggering of
symptoms.

A hypothetical sensitivity distribution for a single symptom for the already chemically
sensitive person in response to a single substance trigger is shown in curve C of Figure 1-1,
and the corresponding dose-response curve is shown in curve C of Figure 1-2. It should be
emphasized, however, that individuals who become chemically sensitive may have been
exposed to an initial priming event that was toxic, as classically defined, and which was the
cause for their having developed chemical sensitivities in the first place.

Conceivably, exposure to certain substances, such as formaldehyde, might elicit all
three types of sensitivities.

The fact that sensitivity means something quite different to toxicologists, allergists, and
clinical ecologists reflects the different disease paradigms under which each operates.
Neither traditional allergists nor toxicologists fully appreciate the two-step process of
induction and triggering that seems to characterize multiple chemical sensitivities.

Those clinical ecologists who reference the literature on classical chemical toxicity to
buttress their case for chemical sensitivity may be adding to the confusion and contributing
to others' reluctance to accept their ideas. Likewise, allergists who dismiss chemical
sensitivity on the grounds that it is not consistent with a recognized immunologic mechanism
may be overlooking another kind of sensitivity in their patients. Although chemicals may act
in some manner (via a toxic mechanism, for instance), to predispose or cause the body to be

reactive to subsequent low-level chemical exposures, the resulting hyperreactivity to low levels of chemically diverse and unrelated substances is not toxicity as classically defined or understood at this time. Some allergists maintain that the term "chemical sensitivity" should not be used in the context we have used it here, but should be reserved only for those responses having an immunological basis. However, the term sensitivity has broader applicability. A parallel might be the word "resistance", which is widely understood whether one is talking about electricity, psychiatry, or an infectious disease. Similarly, "sensitivity" is easily understood when used in any of the three contexts illustrated in this section; it is not the exclusive property of the allergist.

Although chemically sensitive patients were first described by Randolph in the 1950's, the problem seems more prominent in the past decade or so. There are some historical developments which may have contributed to a recent increase in chemical sensitivity.

Americans today spend many more hours per day indoors at work and at home, in schools, shopping malls, and other buildings than preceding generations (Environmental Protection Agency, 1989). On the average, we spend 90 percent of our time indoors. With the concern for energy conservation following the oil embargo of the 1970's, homes and office buildings in the United States were constructed more tightly and make-up air (fresh air intake) was cut to a minimum. Similarly, homeowners and new home builders caulked and sealed, installed storm windows and extra insulation, and effectively reduced fresh air infiltration. On the average, newer homes have half as much fresh air infiltration as older homes (0.5 versus 0.9 air changes per hour) (Mage and Gammage, 1985).

Over the past decade, EPA has conducted TEAM (Total Exposure Assessment Methodology) studies on a variety of volatile organics (1980-1987), carbon monoxide (1982-1983), pesticides (1986-1989), and particulates (1987-present). Samples of 20 volatile organic compounds in the personal (indoor) air, outdoor air, drinking water, and breath of approximately 400 residents of New Jersey, North Carolina, and North Dakota were collected (Wallace, 1987).

Levels of indoor air contaminants were often many times higher than outdoor levels, and sometimes orders of magnitude higher than outdoor levels. Breath levels for most chemicals measured were 30-40 percent of indoor air levels, but measured up to 90 percent of indoor air levels in some cases - tetrachloroethylene, for example. A study of non-occupational pesticide exposure also showed dramatically higher concentrations of pesticides inside homes than out of doors (Immerman, 1990).

Remarkably, the sources of pollutants that were identified by the EPA in homes are the same ones individuals with multiple chemical sensitivities identify as provoking their often vague and seemingly inexplicable symptoms, for example, room air deodorizers, attached garages, hot showers and spas, dry cleaned clothing, cleaning agents and disinfectants.

Before World War II, U.S. production of synthetic organic chemicals totaled fewer than a billion pounds per year. By 1976, production had soared to 163 billion pounds annually (Odell, 1980). Increased sources of indoor air pollution, coupled with decreased fresh make-up air, have transformed the indoor environment. Community exposures to toxic chemicals, industrial and office exposures, and other episodic exposures of individuals also increased, reflecting the rise in production of coal- and oil-derived chemicals and synthetics.

These changes in chemical production, consumer products, and building design have been accompanied by an increasing number of people who appear to react to low levels of environmental pollutants. Interestingly, since World War II certain illnesses, such as asthma

(Sly, 1983) and depression (Klerman and Weissman, 1989), seem to have shown upsurges. It is easy to imagine that asthma could be related to chemical exposures. In the case of depression, it is recognized that solvent-exposed workers experience more depression and cognitive difficulties. Further, the majority of indoor air contaminants are solvents, albeit concentrations are generally much lower than those found in an industrial setting. Randolph often referred to chemical sensitivities as the "petrochemical problem" because the increase in the incidence of this illness seems to parallel the growth of the petrochemical industry and the increased use of synthetic materials such as particleboard, pesticides, synthetic textiles, plastics, and food additives by consumers since World War II.

Randolph, who had been hospitalizing patients and testing them for their food sensitivities, found a critical element in many of his patients' recoveries was avoidance of environmental chemical exposures in their jobs and homes while in the hospital. He developed "comprehensive environmental control", a diagnostic approach in which patients avoid exposure to synthetic chemicals in order to facilitate diagnosis of chemical sensitivity.

Briefly, this technique involves placing the patient in a specially constructed environment devoid of materials that off-gas; avoiding the use of drugs, cosmetics, perfumes, synthetic fabrics, pesticides, and similar substances; and having the patient fast for a period of days until symptoms resolve. This initial period of avoidance and fasting requires approximately 4 to 7 days on the average. During this time, the patient exhibits withdrawal symptoms such as headache, malaise, irritability, or depression. By the end of this time, the patient's symptoms, if environmentally related, should clear, provided that end-organ damage has not occurred. At the end of this avoidance phase, the patient reportedly has a markedly lower pulse rate and an increased sense of well-being, as well as resolution of symptoms. Drinking waters from a variety of sources also are tested to find one most compatible with the patient. Next, individual foods are reintroduced, one per meal, over a two- to three-week period. Following this, the patient is placed on a rotating diet of "safe" foods (i.e., foods that did not provoke symptoms for that particular patient). Finally, the patient is challenged with very low levels (levels routinely encountered in daily living) of common chemicals. Those exposures, both food and chemical, that induce symptoms are to be avoided.

A description of comprehensive environmental control and its role in diagnosis and therapy first appeared 30 years ago in Clinical Physiology (Randolph, 1960) and again in the Annals of Allergy in 1965 (Randolph, 1965).

The detailed design of an environmental unit is beyond the limits of this discussion, however, some of the essentials are noted here. Although conceptually simple and scientifically elegant, achieving a well controlled environment within the average hospital is technically difficult.

First, by employing construction materials, furnishings, and clothing that are less likely to off-gas, very low levels of volatile organic compounds (for example, from synthetics) are maintained inside the unit. To create and operate a unit that is as free as possible from chemical pollution requires knowledge, precision, and vigilance while working with architects, ventilation engineers, contractors and their suppliers, nurses, dieticians, food and water suppliers, and maintenance and custodial staffs.

Several units have been operated by the clinical ecologists and one, which was patterned after those of the ecologists, was operated by John Selner, a Denver allergist. Currently none are in operation, although all of the physicians who have been involved with these units have found them to be a valuable tool for the evaluation of certain patients.

The clinical ecologists' environmental units and Selner's unit shared many of the same design and operational parameters (Table 2). Studies from ecologists' units leave much to be desired in terms of study design. Unfortunately, no studies were ever published from the allergists' unit in Denver.

TABLE 2

Features of Environmental Units[a]

Characteristics/Practices	Allergists' Unit[b]	Clinical Ecologists' Units[c]
Construction using materials that do not off-gas (primarily glass, stell, ceramic; cotton bedding and clothing). Avoidance of synthetic materials. No perfumes, cosmetics, odorous cleaners/ soaps,etc.	Yes	Yes
Air supply filtered; patients' rooms under positive pressure to reduce contamination from adjacent areas; airlocks	Yes	Yes
Patients' medications discontinued insofar as possible; gradual withdrawal from steroids, etc.	Yes	Yes
Patients fasted for 4 to 8 days to clear symptoms.	Yes, if symptoms do not clear after several days in unit	Yes at time of admission to unit
Organic foods used for food testing; commercial foods tested also	Yes	Yes
Patients tested for acceptable water	Yes	Yes
Challenges performed using single foods and chemicals after period of avoidance (to eliminate masking)	Yes	Yes

[a]None of the units described in this table is currently in operation.
[b] Selner in Denver (Selner and Staudenmayer, 1985).
[c]Randolph in Chicago and Rea in Dallas.

By isolating his patients from their usual environments and then re-exposing them to various foods and chemicals one by one, Randolph observed that adaptation seemed to play an important role in his patients' responses to many common substances they ate, drank or inhaled. Adaptation is known in other contexts as "acclimation" or "acclimatization", "habituation", "developing tolerance" and even "addiction". Randolph used the terms "adaptation" and "addiction" most often. However, reference to one of the other words may make it easier to grasp the concept. "Acclimatization" is a widely used term in occupational health that refers to workers gradually becoming accustomed to exposures on the job, for example, heat stress. Understanding adaptation is important for two reasons: (1) adaptation may interfere with the discovery of the effects of a particular exposure on the body and (2) chemical exposures may adversely impact adaptation mechanisms and thus lead to illness.

That human beings respond to chronic exposure to environmental challenges by adapting, acclimating, acclimatizing, or even becoming addicted is widely recognized for a variety of substances. Most would agree that the use of narcotics, alcohol, nicotine, and even caffeine can be addicting. For example, the first cigarette ever smoked might be associated with eye and throat irritation, but over time, with more cigarettes, most individuals adapt, and primarily the pleasurable effects of nicotine on the brain are experienced. After months or years, more cigarettes (or alcohol or caffeine or drugs) may be required for the same amount of lift. The individuals may exhibit addictive behavior, seeking cigarettes more frequently. Subsequently, quitting cigarettes (or alcohol, caffeine, or drugs) may lead to withdrawal symptoms including irritability, drowsiness, fatigue, moodiness, and headache. The reformed smoker may become extremely intolerant of the smoke of others, even in tiny amounts. Suddenly recalled are the irritation and unpleasant feelings associated with the first cigarette ever smoked. Over time the individual had "adapted" to those effects. Adaptation, which on the surface would seem good for the organism, may in fact be a two-edged sword. Developing tolerance for the noxious properties of the exposure may allow the individual to remain in the exposure more comfortably while other harmful consequences of the exposure continue. Thus the heavy smoker who is "adapted" to tobacco smoke is at increased risk for developing emphysema, lung cancer or vascular disease. While often occurring at much lower levels of exposure than the above examples, food and chemical adaptation and addiction have been observed by some physicians in their patients. In the case of MCS patients, *multiple* incitants, not only tobacco smoke, may be involved and *all* may need be avoided simultaneously for improvement to occur. Thus, frequent exposure to a substance results in adaptation (irritation and other warning signals may disappear). Continued exposure may lead to addiction. Reduction or cessation of exposure generally results in withdrawal symptoms.

What may confuse patients and practitioners is that the symptoms for which the individual is most likely to seek a physician's help are those that occur during withdrawal when the person is no longer exposed (or is less exposed) to the offending agent. Thus headaches may occur when the individual smokes fewer cigarettes than usual or drinks less caffeine. Indeed, these unpleasant withdrawal symptoms may be forestalled by smoking another cigarette or taking another drink of coffee, thus perpetuating addiction. Patients may report that smoking a cigarette or drinking a cup of coffee in the morning (after 8 or so hours without) relieves their headache (a withdrawal symptom) and they feel better, not suspecting that the cigarette or coffee might also be the cause of their headache.

Occupational health presents many examples in which acclimatization, inurement, or tolerance to a substance is known to develop, for example, exposure to ozone, nitroglycerin, and solvents. Note that the incitants mentioned thus far are all quite different from one another: some are ingestants, others inhalants; some are solid, others liquid or gaseous in form; some are simple molecules, whereas others are complex mixtures. The point is that the human body appears able to adapt to an endless array of substances.

By isolating MCS patients from their usual environments and then re-exposing them to various foods and chemicals one by one, physicians have observed that many common substances patients eat, drink or inhale seem to provoke symptoms.

A biphasic response to some of these substances (Figure 2) has been reported. Initially the individual might experience a stimulatory effect (adapted response; tolerance develops) lasting varying periods of time depending upon the incitant. However, this "up" phase was generally followed by a withdrawal phase (maladapted response; loss of tolerance). Upon beginning to experience unpleasant withdrawal symptoms, the individual would seek, consciously or unconsciously, more of the same substance. These ups and downs follow a sort of sinusoidal (biphasic) pattern, as depicted in Figure 2. On the graph, beginning at zero, the patient is free of symptoms and at baseline health status. Following a one-time or occasional exposure to a provoking substance, stimulatory effects result; after a period of time (minutes to hours to days, depending upon the nature of the incitant), the stimulatory effects subside and give way to withdrawal symptoms. The frequency of these up and down reactions depends upon the frequency of exposures, and the amplitude of the stimulatory and withdrawal portions of the reaction depend upon the substance and the individual's susceptibility (degree of adaptation or addiction) to it. The particularly sensitive person exhibits larger amplitudes than the normals. The key to understanding multiple

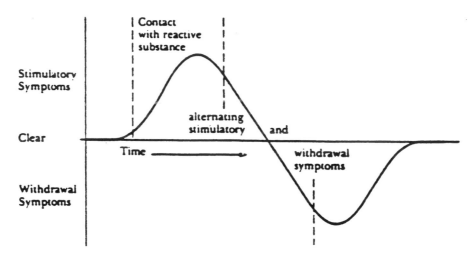

FIGURE 2   Symptom progression of a single reaction to an incitant. During the early phases of exposure to a particular substance, stimulatory symptoms predominate ("up," "hyper," "jittery"). As exposure to the offending agent continues, adaptation occurs and fewer of these symptoms are experienced. With removal from (or discontinuance of) exposure, the individual experiences withdrawal symptoms ranging in intensity from mild to severe. (From O'Banion, D.R., Ecological and Nutritional Treatment of Health Disorders, 1981, p. 68. Courtesy of Charles C. Thomas, Publisher, Springfield, Illinois.)

chemical sensitivity may lie in recognizing these ups and downs that appear to occur after exposure to many different substances. The amplitude of a reaction varies from person to person and incitant to incitant, but the pattern is reported to be quite constant.

After long-term exposure to a given incitant (for instance, alcohol), especially in certain sensitive individuals, the degree and duration of stimulation may become less and less while the withdrawal or depressed phase becomes deeper and more prolonged. At face value, this sinusoidal reaction to a substance might seem a somewhat artificial construct, but Randolph asserts it is not.

Chemical sensitivities may be difficult to assess while a patient remains at home or even in most hospitals because these places generally contain background low levels of natural gas, disinfectants, perfumes, cleaners, tobacco smoke, paints, varnishes, adhesives, and other substances. The patient's symptoms may be masked by the presence of these contaminants.

Under normal living circumstances, the stimulatory and withdrawal levels for foods and chemicals overlap each other (Figure 3) so that in real life - outside an environmental unit - at any given moment what the organism may be feeling is a summation of all effects, whether stimulatory or depressive, of all substances recently inhaled, contacted, or ingested. Figure 3 illustrates that attempts to identify the effects of single substances would be frustrated by the overlapping responses. Only by placing the individual in an environment devoid of chemical and food incitants would one be able to determine whether the illness is alleviated. Assuming the patient improves (which occurs in the majority of cases, according to ecologists), the next step would be to reexpose the person to individual substances in order to avoid overlapping responses, and then to observe the result. If all possible food and chemical contributors are not removed, an effect may be missed. Hence, in order to rule out environmental illness definitively, an environmental unit would be required. Conceivably environmental illness could be ruled in on an outpatient basis, but not ruled out.

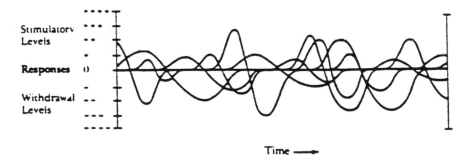

FIGURE 3   Overlapping of responses to food and chemical incitants in an individual with multiple exposures and multiple chemical sensitivities.

In real life, stimulatory and withdrawal reactions are observed but often not understood. For example, an asthmatic might feel well after spending a week on a Carribean island, breathing relatively uncontaminated air and eating a diet devoid of usual foods, only to have a severe, life-threatening asthmatic response to exhaust from the engine of a boat taking the individual home. Once back home in a metropolis, the asthmatic readapts, acclimatizes to auto exhaust, combustion products and other air pollutants in the area, and experiences only chronic wheezing. Thus, following deadaptation (removal from incitants), the individual exhibits a more acute and convincing reaction upon reexposure. This appears to be what occurs in an environmental unit during testing. So acute and convincing are some of these reactions that patients themselves erroneously (at least in the eyes of some) surmise they must have an "allergy" to a particular substance. However, if the patient is not deadapted (unmasked) when tested, a reaction may not occur, convincing the physician that the "allergy" was all in the patient's mind.

Occupational health has several widely recognized examples of adaptation that are analogous (Ashford and Miller 1991). They, too, fit a biphasic pattern. Industrial hygienists and occupation health physicians know that one of the most valuable clues to work-related illness is a history of intense symptoms following return to work after a vacation or weekend (leading to withdrawal and deadaptation).

Ozone, an air pollutant of special concern to residents of Los Angeles and other cities, has been the focus of considerable research relevant to adaptation. Intrigued by how little respiratory illness and death occurred relative to the high levels of ozone in very polluted cities and suspecting adaptation might play a protective role, Hackney and associates (Hackney et al., 1977a) compared the responses of four Canadians (not adapted) and four Californians (adapted) to ozone challenges. Although reactivity varied greatly from individual to individual, Californians were only minimally reactive to levels that for the Canadians caused coughing, substernal discomfort and airway irritation, pulmonary function test decrements, and increased red blood cell fragility.

In another experiment, six volunteers with respiratory hyperreactivity were placed in an environmental chamber with ozone at 0.5 ppm (parts per million), typical of high ambient levels, for 4 days (Hackney et al., 1977b). Five of six had decreased pulmonary function during days 1 to 3, but gradually improved almost to baseline by day 4, suggesting adaptation had occurred. The authors note that not all adverse effects of ozone may be prevented by adaptation; for example, increased red blood cell fragility may persist. Therefore, adaptation or masking of some symptoms may occur while other physiological alterations continue.

Individuals' abilities to adapt to ozone appear to depend upon their initial sensitivity to it. More sensitive persons adapt more slowly and cannot maintain the adaptation as long; they usually remain adapted less than 7 days following cessation of exposure (Horvath, 1981). While nitroglycerin and ozone adaptation (and deadaptation) may differ in certain respects from the adaptation (and deadaptation) described in MCS patients, solvents are among the chemicals most frequently implicated by chemically sensitive patients who attribute the onset of their illness to a particular exposure (Terr, 1989; Cone et al., 1987) and adaptation to solvents has also been documented. Vapors from various solvents are the most prevalent of indoor air contaminants (Molhave, 1982). The volatile organic compounds (VOCs) associated with sick building syndrome are in large part solvent vapors. The sensory irritation, headache, drowsiness, and other symptoms noted by occupants of tight buildings are consistent with known effects of solvent vapors, albeit at much higher concentrations.

Those who have painted or used solvents to any major extent are well aware of the olfactory fatigue (nasal adaptation) that occurs and may have experienced the stimulatory and depressive properties of solvents. Alcoholic beverages contain the solvent ethanol, which has related and familiar stimulatory and withdrawal effects.

Studies of xylene, one of the most prevalent solvents in indoor air, demonstrate that its effects are attenuated as exposure continues, presumably due to adaptation (Riihimaki and

Savolainen, 1980). Riihimaki and Savolainen exposed healthy male volunteers to constant (100 or 200 ppm) and varying (200 or 400 ppm hourly peak) concentrations of xylene, adjusting baseline concentrations in the latter case so that a mean concentration of 100 or 200 ppm was maintained. Exposures occurred over a six-hour period (with a one-hour break at noon) for five days, followed by a two-day weekend and one to three more days of active exposure to xylene. A variety of psychophysiologic parameters were measured, including reaction time, body balance, manual dexterity, and nystagmus.

Of particular interest, Riihimaki and Savolainen (1980) observed that most of the adverse effects of xylene upon their normal subjects "tended to disappear after a few succeeding days of exposure." However, "after the weekend away from exposure, the effects were again discernible." They conclude: "This phenomenon suggests that tolerance had developed over a few days with regard to psychophysiological effects by xylene."

With regard to patients with chemical sensitivities who also develop dietary intolerances, Bell notes that "foods are not only sources of nutrients, but also complex mixtures of organic chemicals. For instance, it is the unique pattern of chemical constituents that make a tomato a tomato rather than an apple" (Bell, 1982, pp. 35-36). Interestingly, limonene and pinene which are present in oranges also are constituents of room air deodorizers which provoke symptoms in some chemically sensitive patients. Like airborne pollutants, foods contain a wide range of chemical constituents and are in intimate contact with the organism for long periods of time. The surface area of the gastrointestinal tract is enormous, and the chemical load, in terms of both quantity and diversity of exposure, is huge.

We have mentioned a number of exposures that are recognized as involving adaptation. What is clear is that individuals with or without multiple chemical sensitivities undergo adaptation to a wide variety of substances in their environment. What is not clear is the specific role adaptation plays in the dramatic responses patients with food and chemical sensitivities have to low-level exposures that do not overtly affect others. These concepts are familiar to occupational health practitioners and industrial hygienists because they observe such effects firsthand among workers exposed to chemicals. Randolph states that most physicians see patients long after adaptation has occurred and at the time when end organ damage is setting in: "It is much as if the physician arrived at the theatre sometime during the last scene of the second act of a three act play--puzzled by what may have happened previously to the principal actor, his patient" (Randolph, 1962, p. 7). Through comprehensive environmental control (that is, an environmental unit), one may be able to overcome the masking effect of adaptation and back up or reverse the exposure to allow monitoring of toxicity in progress. The environmental unit may represent a kind of *dynamic toxicology*; traditional medical approaches provide only a snapshot of what is happening to the patient.

There are several reasons why de-adapting patients is critical to the study and diagnosis of MCS:

1. People are often exposed to dozens of different incitants simultaneously (such as volatile organic compounds in a tight home or building) and literally hundreds of different incitants over the course of a single day, so that health effects of these exposures may overlap, making it difficult to discern cause-and-effect relationships.

2. With continuous or frequent exposure to the same substance or chemically-related substances (such as xanthines in coffee, tea, chocolate and colas), individuals adapt or, in other words, develop tolerance to those exposures. Acute symptoms gradually may give way to chronic symptoms that bear no apparent

relationship to any particular exposure. Exposures may never stop long enough for the patient to reach baseline.

3. Exposures that are initially pleasant or stimulating (such as alcohol, solvents, or nicotine) generally also have withdrawal effects such as headache, depression or irritability, associated with them. Such withdrawal symptoms may occur hours to a few days after cessation of, or reduction in, exposure, greatly confounding attempts by patients and physicians to relate symptoms to a particular incitant.

Comprehensive environmental control, that is, use of an environmental unit, can overcome the masking effect of adaptation and the problems of overlapping exposures that result in overlapping responses to multiple agents. The environmental unit can back up or reverse the experience of adaptation and allow the investigator to monitor toxicity in progress. Figure 4 graphically depicts the changes in symptoms that might occur in a patient after entering an environmental unit. The advantages dynamic toxicology of this nature has over conventional methods for determining toxicity include facilitating detection of subclinical, prepathological effects of chemicals and providing more than just a snapshot of an individual's response to substances. Removing the person from interacting, time-dependent stimuli in this way allows the unraveling of multiple causes. The environmental unit is an essential tool. Many carefully conducted studies of chemical effects that have had negative or equivocal outcomes in the past may have been flawed by their failure to take adaptive mechanisms into account. The potential consequences of such an oversight are major.

Entry Into   Challenges
Environmental Unit   Begin

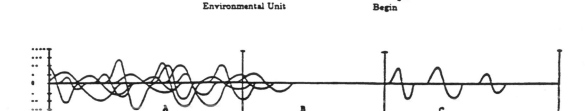

FIGURE 4   Graphical representation of an individuals symptoms before and after entering an environmental unit. In time period A an individual is responding to multiple incitants encountered in normal daily living (chemicals and/or foods), with stimulatory and withdrawal effects that overlap in time. At any particular time, how the person feels is determined not only by ongoing exposures, but by previous exposures whose effects may still be waning. In time period B, the individual enters an environmentally controlled facility, fasting. With cessation of contributory exposures, withdrawal effects occur, for example, headache, fatigue, and myalgias. Symptoms continue for some time (typically for 4-7 days) until the individual reaches "0" baseline. In time period C, single challenges to suspected incitants are administered. Symptoms, often robust, develop soon after challenges, allowing patient and physician to begin to observe the cause-and-effect relationship between exposures and symptoms for that individual.

Important questions that must be addressed in future studies of chemical sensitivity include:

1. Are subjects in a deadapted state prior to challenges so that extraneous

exposures during and prior to a challenge (up to several days before) do not interfere with testing?

2.   Are open challenges performed first to confirm that the placebo (clean air or a masking odor such as peppermint) is in fact a placebo and that the "active" challenge is something to which the patient has had demonstrable reactions?

3.   What is the recency and latency of the patient's exposure to the substance being tested? In other words, has enough time elapsed (about a week or so) that the person is no longer adapted or reacting to the last exposure but not so much time that the sensitivity has waned? Recency of exposure is recognized as a crucial variable in conducting challenges in patients with occupational asthma, for example.

The rift between allergy and clinical ecology has been fueled by the difficulty inherent in communicating these complex observations concerning adaptation, with unfortunate consequences for patients. An ancient proverb observes "When elephants fight, it is the grass that suffers". When physicians are embattled, it is the patient who suffers. Carefully designed studies of deadapted patients in an environmental unit, using double-blind placebo-controlled challenges, are an essential first step for helping resolve current professional antagonisms and placing research in this field on scientific footing.

SUMMARY OF ADAPTATION HYPOTHESES

Symptoms of exposure to many chemicals, whether inhaled or ingested, appear to follow a biphasic pattern. Adaptation is characterized by acclimatization (habituation, tolerance) with repeated exposures that result in a masking of symptoms. Withdrawal occurs when exposure is discontinued. Once a person has adapted, then the experimental consequences are that further exposures have very little additional effect and therefore may not be observed. The observer may not be able to witness the stimulatory or reactive event because a kind of "saturation" effect has set in.

Adaptation and withdrawal occur for a wide variety of organic and inorganic substances in many physical forms, including various dusts and fumes, solvents, nitroglycerin, ozone, drugs and foods.

An individual is exposed to a variety of substances at different times with varying frequency, duration, and intensity of exposure for each of these substances and with varying frequency and duration of reduction in or cessation of exposure for each substance. The individual may be in different stages (stimulatory or withdrawal) simultaneously for different substances. These stages may overlap and interfere with attempts to observe cause-and-effect relationships.

Adaptation may mask some symptoms or effects while other physiological alterations may continue.

Comprehensive environmental control, that is, use of an environmental unit, can overcome the masking effect of adaptation and the problems of overlapping exposures that result in overlapping responses to multiple agents. The environmental unit can back up or reverse the experience of adaptation and allow the investigator to monitor toxicity in progress. The advantages dynamic toxicology of this nature has over conventional methods for determining toxicity include facilitating detection of subclinical, prepathological effects of chemicals and providing more than just a snapshot of an individual's response to substances. Removing the person from interacting, time-dependent stimuli in this way allows the unraveling of multiple causes.

REFERENCES

Ashford, N. A. and Miller, C. S. 1991. Chemical Exposures: Low Levels and High Stakes. New York: Van Nostrand Reinhold.

Bell, I. R. 1982. Clinical Ecology. Bolinas: Common Knowledge Press.

Cone, J. E., Harrison, R., and Reiter, R. 1987. Patients with multiple chemical sensitivities: clinical diagnostic subsets among an occupational health clinic population. Philadelphia: Hanley & Belfus.

Environmental Protection Agency, Office of Air and Radiation. 1989. Report to Congress on Indoor Air Quality.

Hackney, J. D., Karuza S. K., Linn, W. S. 1977a. Effects of ozone exposure in Canadians and southern Californians, evidence for adaptation? Arch. Environ. Health. 32:110-116.

Hackney, J. D., Linn, W. S., Mohler, J. G., and Collier, C. R. 1977b. Adaptation to short-term respiratory effects of ozone in men exposed repeatedly. J. Appl. Psychol. 43:82-85.

Horvath, S. M., Gliner, J. A. and Folinsbee, L. J. 1981. Adaptation to ozone: duration of effect. Am. Rev. Respir. Dis. 123:496-499.

Johnson, A. and Rea, W. J. 1989. Review of 200 cases in the environmental control unit, Dallas. Presented at the Seventh International Symposium on Man and His Environment in Health and Disease, February 25-26, Dallas, Texas.

Immerman, F. and Schaum, J. 1990. Final Report of the Nonoccupational Pesticide Exposure Study. U.S. EPA, Research Triangle Park.

Klerman, G. L., and Weissman, M. M. 1989. Increasing rates of depression. J. Amer. Med. Assoc. 261: 2229-2235.

Mage, D., and Gammage, R. B. 1985. Evaluation of changes in indoor air quality occurring over the past several decades. In Gammage, R. B. and Kaye, S. (editors). Indoor Air and Human Health. Chelsea: Lewis Publishers.

Molhave, L. 1982. Indoor air pollution due to organic gases and vapors of solvents in building materials. Environ. Internat. 8:117-127.

National Foundation for the Chemically Hypersensitive. 1989. Cheers. 1:6.

O'Banion, D. R. 1981. Ecological and Nutritional Treatments of Health Disorders. Springfield, IL: Charles C. Thomas.

Odell, R., Environmental Awakening. 1980. Cambridge, MA: Ballinger.

Randolph, T. G. 1960. A third dimension of the medical investigation. Clin. Physiology. 2(1):42-47.

Randolph, T. G. 1962. Human Ecology and Susceptibility to the Chemical Environment. Springfield, Illinois: Charles C. Thomas.

Randolph, T. G. 1965. Ecologic orientation in medicine: comprehensive environmental control in diagnosis and therapy. Ann. Allergy 23:7-22.

Randolph, T. G. 1987. Environmental Medicine: Beginnings and Bibliographies of Clinical Ecology. Fort Collins, Colorado: Ecology Publications.

Randolph, T. G. and Moss, R. W. 1989. An Alternative Approach to Allergies. New York: Harper & Row.

Riihimaki, V. and Savolainen, H. 1980. Human exposure to m-Xylene. Kinetics and acute effects on the central nervous system. Ann. Occup. Hyg. 23:411-432.

Selner, J. C. and Staudenmayer, H. 1985. The relationship of the environment and food to allergic and psychiatric illness. Pp. 102-145 in Psychobiology of Allergic Disorders. Young, S. and Rubin, J. editors. New York: Praeger.

Sly, R. M. 1988. Mortality from asthma. J. Allergy Clin. Immunol. 82(5):705-717.

Terr, A. I. 1989. Clinical ecology in the workplace. J. Occupat. Med. 31(3):257-261.

Wallace, L. A., Pellizari, E. D., Hartwell, T. D., Sparacino, C., Whitmore, R., Sheldon, L., Zelon, H. and Perritt, R. 1987. The TEAM study: personal exposures to toxic substances in air, drinking water, and breath of 400 residents of New Jersey, North Carolina, and North Dakota. Environ. Res. 43:290-307.

Design of Animal Models to Probe The Mechanisms of Multiple Chemical Sensitivity

Meryl H. Karol

INTRODUCTION

Recent years have witnessed the emergence of a syndrome called multiple chemical sensitivity (MCS) characterized by recurrent symptoms related to multiple organ systems (1). Symptoms associated with the disorder include: wheeze, cough, shortness of breath, as well as headache, fatigue, depression, confusion, muscle ache, weakness, and gastro-intestinal upset (2). The ailment has been associated with diverse environmental exposures, most frequently to volatile organic solvents (2).

As with many newly recognized medical syndromes, development of an animal model has been suggested as an approach for gaining increased understanding of the pathogenesis of the disorder. Critical design of an animal model is essential to permit both acquisition of meaningful information relative to disease onset and pathogenesis, and to allow scientific acceptance of the model. With MCS, however, the broad range of symptoms, together with their subjective nature, makes it difficult to design an appropriate animal model.

The alternative approach frequently used for development of animal models, simulation of causal exposure conditions, is also fraught with difficulty. Exposures associated with onset of the syndrome are poorly defined and typically involve mixtures of chemicals, glues, solvents, and pesticides (2).

Design of an animal model for pulmonary disease resulting from environmental exposures will be discussed and emphasis placed on features of the model which have application to MCS.

ANIMAL MODELS OF ENVIRONMENTAL LUNG DISEASES

Animal models have been developed for numerous environmental lung diseases including: allergic hypersensitivity (3), acute (4) and chronic byssinosis (5), organic disease toxic syndrome (6), sensory irritation (7), as well as for chronic bronchitis, emphysema and lung cancer (8). Many have proved beneficial in elucidation of: disease pathogenesis (8), progression (5) or in identification of biomakers associated with chemical exposures (9).The characteristics of an ideal animal model have been summarized (Table 1). The use of small animals wherever possible is both economically beneficial and fosters use of the lowest possible phylogenetic species. Exposure of the model should resemble as closely as possible that of the human, to enable distribution of the chemical xenobiotic to appropriate target organs. Lastly, endpoints of toxicity must be carefully selected to resemble, or have relevance to, those observed in humans. Endpoints should occur with low spontaneous incidence.

TABLE 1

Features of an Ideal Animal Model for Environmental Disease

— Use of a small animal species
— Exposure through a natural route
— Similarity in site of lesion development in humans
— Similarity in cell type affected
— Low incidence of toxicity end point occurrence in controls
— Similar biological handling of toxicant (absorption, metabolism, storage, excretion) to humans

Modified from ref. 8.

Attention to design in development of an animal model for pulmonary sensitivity (10) has allowed its use in study of several environmentally related lung disorders. The basic features allowing for exposures and monitoring of responses are presented in Figure 1 and listed in Table 2. Guinea pigs are used since they are readily accessible, small rodents known to demonstrate strong allergic anaphylactic lung responses. Exposure is by inhalation of airborne atmospheres of chemical xenobiotics. Chemical atmospheres are generated in the mixing chamber. The concentration of xenobiotic, as well as the size of aerosol particles is established from samples taken at the ports in the mixing chamber. The desired chemical concentrations are achieved by adjusting the generator feed and airflow exhaust from this chamber.

Animals are exposed to this atmosphere by connecting individual plethysmographs to the mixing chamber using polytetrafluoroethylene (PTFE) tubing. The system is designed to accommodate 4 plethysmographs simultaneously.

Although exposure is predominantly via the inhalation route, dermal contact with the xenobiotic also occurs. Further, through grooming maneuvers and operation of the respiratory escalator, ingestion of the xenobiotic also occurs. Additionally, deliberate exposure through the ingestion route can be provided in this model by incorporating xenobiotics into the food supply placed within each plethysmograph.

During exposure, animals are neither sedated nor restrained within plethysmographs. The benefit derived from this design is the maintenance of a normal body position and therefore an appropriate breathing pattern of animals. This results in the deposition of inhaled particles in the appropriate regions of the respiratory tract and thereby allows

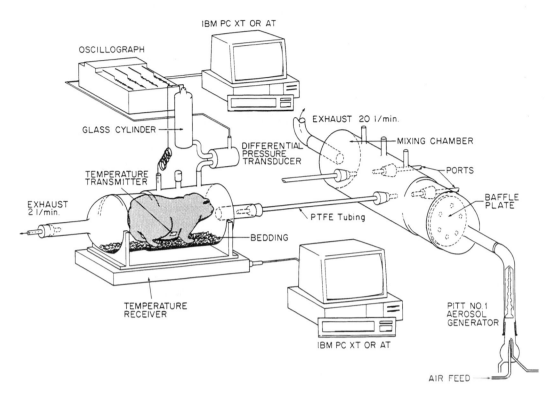

Figure 1   Schematic presentation of the exposure system and equipment for monitoring of pulmonary and febrile responses to airborne xenobiotics. (From ref. 11 with permission).

TABLE 2

Features of an Animal Model for Pulmonary Sensitivity

— Use of guinea pigs (250-400g)
— Exposure by inhalation of airborne chemicals
— Use of unsedated and unrestrained animals
— Avoidance of immunologic adjuvants to achieve sensitization
— Passive recording of responses without disturbance of animals
— Continuous monitoring of pulmonary and febrile responses for 24 hours

suitable biologic handling of xenobiotics (see Table 1).

A second feature of the model is the avoidance of adjuvants to induce immunologic sensitization and of physical restraint within the chamber.

Exposure duration may extend from seconds to hours since atmospheres are continuously created using an aerosol or dust (11) generator.

Monitoring of Responses

An essential feature of this model is the provision for passive monitoring of responses. Continuous recording of breathing pattern is achieved through the use of pressure transducers. Core temperature is monitored using radio frequency transmitters and receivers (see Figure 1). Fever has been associated with several pulmonary syndromes including hypersensitivity pneumonitis and organic dust toxic syndrome.

All readings are collected electronically without disturbance of animals (12), and recorded continuously for up to 24 hrs. The latter feature permits detection of reactions, such as breathing difficulty and fever, which have a variable time of onset. This feature was paramount to identification of late-onset airway hypersensitivity responses to ovalbumin (13) and tuberculin protein (14). The model is applicable for study of illness thought to result from environmental contact.

EVALUATION OF ANIMAL MODELS

Once developed, an animal model must be evaluated. Ideally, the syndrome elicited in the model will closely resemble that reported for the human experience. Such situations generate confidence that mechanisms of pathogenesis in the animal system will have relevance for extrapolation to human disease pathogenesis. The guinea pig model for environmental lung disease has been evaluated for three clinical syndromes as described below.

Respiratory Hypersensitivity

Respiratory hypersensitivity to aeroallergens is characterized by occurrence of immediate, or late-onset airway disturbance (15). Evaluation of the model has been performed by comparison of responses in the guinea pig with those reported for clinical sensitization. Characteristic features of clinical sensitivity are occurrence of early and late onset falls in FEV1 and eosinophilic inflammation of the lung (15). The guinea pig model of pulmonary sensitivity demonstrated each of these responses (12). Animals responded to inhalation provocation challenge with an immediate airway constrictive response (see Figure 2). Recovery was complete within one hour. Continuous monitoring of the breathing pattern revealed a second response with maximum intensity occurring about 5 hours following provocation challenge. This late-onset response resolved over several hours.

At the height of response, histopathologic examination of pulmonary tissue revealed eosinophilic infiltration of the submucosal and epithelial regions of both the large and small airways (12, 13). Since eosinophilic inflammation is a primary characteristic of allergic pulmonary sensitivity, the animal system demonstrated the major physiological and histopathological features of the human syndrome it was designed to model.

The animal model of allergic sensitivity has been additionally evaluated by comparison of immunologic findings in the guinea pigs with those from clinical patients. In experimental studies of TDI hypersensitivity, antibodies produced in guinea pigs were shown to have specificity towards 2,4 and 2,6 TDI isomers (16). Further, the antibody titer was found to vary directly with the concentration of TDI to which animals had been exposed

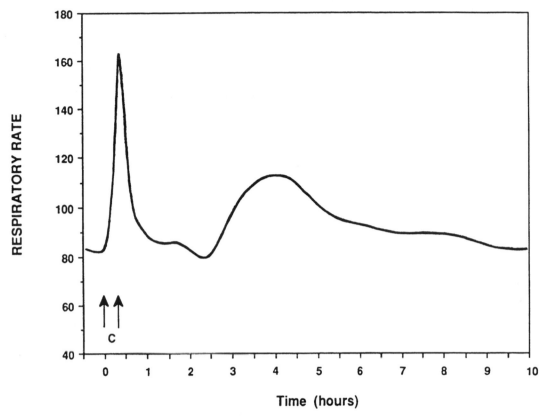

Figure 2   Immediate- and late-onset pulmonary responses of a sensitized guinea pig to ovalbumin. Arrows indicate the period of inhalation challenge. The immediate response occurred within 21 minutes of inhalation challenge. The late-onset response was observed 3.5-4.5 from the start of the challenge exposure. (from ref. 12 with permission).

(17). Both the antibody specificity and its concentration-dependent production were found in sera from individuals occupationally exposed to TDI (16,18).

Guinea Pig Model of Airway Hyperreactivity

One of the cardinal features of asthma is the presence of airway hyperreactivity (AHR) (19). This condition is most frequently assessed as a response to a lower than normal amount of inhaled histamine or methacholine. Study of the pathogenesis of asthma requires methodology for detecting occurrence of AHR. The basic guinea pig model for environmental lung disease was applied to detection of AHR. Methodology was developed to quantify airway reactivity in animals based upon their airway response to increasing concentrations of inhaled histamine (20).

To validate the model, it was necessary to assess its sensitivity in detecting transient AHR associated with an immunologic hypersensitivity response. Accordingly, a study was undertaken to determine whether the newly developed methodology could detect AHR which had been established using conventional methodology (21). Data presented in Figure 3 indicate the occurrence of AHR in each of the experimental animals following the hypersensitivity episode (22).

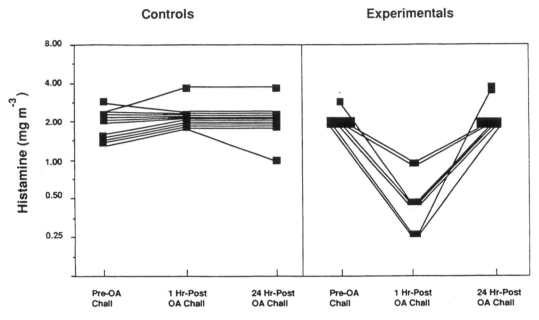

Figure 3   Airway reactivity of control and ovalbumin (OA)- guinea pigs prior to and following inhalation challenge with antigen. Hyperreactivity of experimental animals was apparent 1 hr post OA challenge by the response to lower than normal concentrations of histamine. (from ref. 22 with permission).

The model was further evaluated by comparing its sensitivity for detecting AHR with that of the previously employed methodology. The sensitivity of the two methods appeared comparable. From measurement of dynamic lung compliance and tidal volume (21), reactivity increased from 2.5 to 0.75 mgm-3 histamine required to evoke response. Using the non-invasive methodology (22), reactivity increased from 2.10 mgm-3 histamine to 0.5 mgm-3 required for response. The non-invasive methodology had the additional advantage of accommodating repeated measurements over a prolonged period of time as would be necessary for study of a chronic disease such as asthma.

ANIMAL MODEL FOR ORGANIC DUST TOXIC SYNDROME

A syndrome characterized by fever and neutrophilic alveolitis has been described (23). The disease is associated with environmental exposure to organic dusts and accordingly has been named Organic Dust Toxic Syndrome (ODTS). In order to investigate both the conditions conducive to development of the syndrome, and mechanisms underlying the neutrophilic inflammation, an animal model was sought.

The basic guinea pig inhalation model had features desirable for application to ODTS (Table 2). These features included: capability to provide prolonged exposures via the inhalation route, and to monitor pulmonary and febrile responses which might develop over several hours. In addition, bronchoalveolar lavage could be readily performed in guinea pigs for the evaluation of neutrophilic pulmonary inflammation.

The model was adapted to study of ODTS by exposing animals for 6 hr to atmospheres of cotton dust obtained from a cotton mill in Memphis, TN (6). Monitoring of animals revealed development of the cardinal features of the syndrome, ie., fever, and neutrophilia (Figures 4,5).

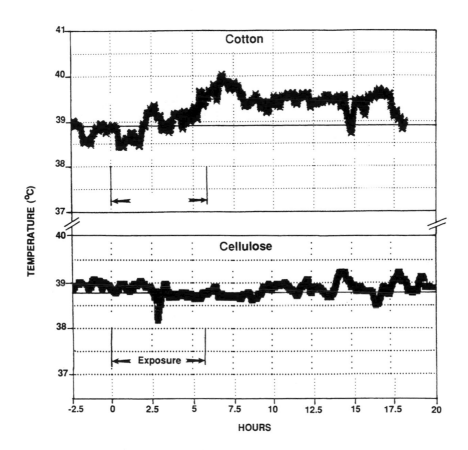

Figure 4   Febrile response of a guinea pig upon inhalation of cotton dust (upper) and lack of response to cellulose dust. Exposure period is indicated by arrows. The febrile response (1.1 oC rise in temperature) was sustained for several hours. (from ref. 6 with permission).

Assessment of the model was achieved by further experiments using "control" dusts. Frequently, in developing animal models, high dosages (or concentrations) of xenobiotics are employed. The rationale for this procedure is based on the fundamental dose-response

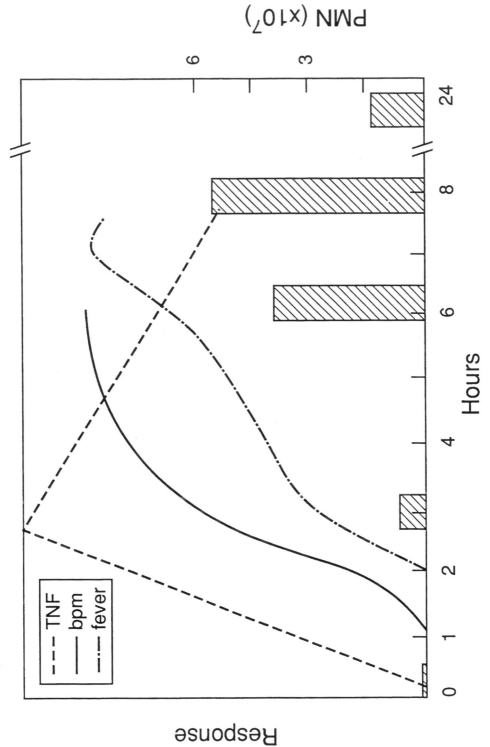

Figure 5 Physiological changes in guinea pigs resulting from inhalation of cotton dust. Hatched bars indicate polymorphonuclear (PMN) leukocyte count (right axis). Left axis: tumor necrosis factor (TNF) activity, breathing frequency, and fever response all shown relative to corresponding pre-exposure values. (from ref. 6 with permission).

principle of toxicology, ie., that if an agent is responsible for an effect, high concentrations of agent will produce sizable responses which can be readily measured. However, high concentrations of agents may create unnatural effects, such as those resulting from overload of biologic systems. For example, inhalation of high concentrations of particulates produced dust overload in rats, resulting in reduced alveolar clearance and altered particle retention (24). The possible occurrence of such effects must be avoided in design of animal models. Experiments must always be performed however, to test for their existence.

Validation of the model for ODTS was achieved by assessing the response of guinea pigs to a high concentration of cellulose dust (22), a dust without reported adverse health effects. Cellulose was selected since it represents the structural matrix of cotton but is devoid of the chemical and biological contaminants usually associated with cotton dust. The absence of both a febrile response, (see Figure 4) and evidence of alveolar inflammation in animals exposed to 41 mgm-3 cellulose, indicated that the response previously detected with cotton dust was not a result of dust overload in the lungs. Rather, it represented pulmonary toxicity resulting from the cotton dust.

AN ANIMAL MODEL OF MCS

Acceptance of an animal model is dependent upon how well the model reproduces the symptom complex associated with the particular disorder. With MCS, the symptoms are highly variable, involving many organ systems. Acceptance of a model for MCS will likely be difficult since there are no objective criteria which are accepted as defining the disorder. Further, the "adaptation" phenomenon, ie., unresponsiveness to acute chemical exposure, implies that negative findings are to be expected and occurrence of symptoms will be unpredictable.

The design of the animal model described above is compatible with study of chronic disease since animals are not instrumented and measurements (such as respiratory patterns and core temperature) are obtained passively and continuously. In addition, blood samples can be obtained readily for immunoglobulin and lymphocyte assessments. RAST (radioallergosorbent) testing can be done to ascertain food intolerance. The model also allows assessment of hypersensitivity, either by skin testing for cutaneous sensitivity, or through provocative inhalation challenge with aeroallergens.

Validation of a model for MCS will be difficult. Which symptoms will be accepted as representing the disorder? Will the absence of immunologic or pulmonary function changes signify a valid consequence of exposure, or an inadequacy of the animal model? Thus, although the animal model described here has many features which make it attractive for use in study of MCS, the absence of responses in animals would be hard to interpret. "Adaptation" could be invoked as a possible reason for unresponsiveness.

A more productive approach for development of an animal model for MCS would appear to be experimental reproduction of exposure conditions associated with development of the disorder. Continuous assessment could be made of physiologic changes (for example, changes in respiratory, neurologic, behavioral or immunologic function). The changes could then be assessed in control animals treated identically to those exposed, but without xenobiotic presence.

It should be noted that reproduction of causative exposure conditions of MCS presents a considerable difficulty since exposures associated with development of the disorder are not well defined. Concentrations of chemicals have not been reported. Moreover, implicated atmospheres are typically variable and complex. One novel solution to this problem would be to introduce animals into the actual environment associated with MCS and to monitor reactions of the animals. However, the uncharacterized and unpredictable nature of exposures associated with development of MCS makes this a difficult undertaking. Simulation of "contaminated" environments appears a more practical yet improbable alternative.

SUMMARY

MCS is a newly recognized disorder associated with exposure to environmental chemicals. Symptoms associated with it are highly diverse and involve many organ systems. Development of an animal model to study the pathogenesis of the disorder is a laudable goal but standard approaches appear implausible. The difficulty lies both with the absence of a clear definition of the disorder and with lack of diagnostic criteria. Further, phenomena associated with the syndrome, such as adaptation, implies that absence of symptoms may be interpreted as a representative "response".

Approaches to animal model development typically focus on either reproduction of responses or of exposures associated with the particular disorder. For MCS, both approaches present difficulties since there is an absence of agreement on diagnostic criteria and little information on agents, lengths of exposure, or number of repeated exposures necessary for development of the disorder.

Animal models have been developed for many environmentally related disorders and have provided information essential to elucidation of disease pathogenesis, diagnosis and treatment. It is hoped that this approach can be applied to study of MCS since much remains to be learned about this very puzzling but increasingly reported disorder.

ACKNOWLEDGEMENT

Support from NIEHS 01532 is gratefully acknowledged.

REFERENCES

1.   Cullen, M.R. The worker with multiple chemical sensitivity: an overview. In: Workers with Multiple Chemical Sensitivity. State of the Art Reviews, Occupational Medicine (M.R. Cullen, Ed.) Hanley and Belfus, Inc., Philadelphia, PA, pp. 655-662 (1987).

2.   Cone, J.E., Harrison, R. and Reiter, R. Patients with multiple chemical sensitivities: Clinical diagnostic subsets among an occupational health clinic population . In: Workers with Multiple Chemical Sensitivity. State of the Art Reviews, Occupational Medicine (M.R. Cullen, Ed.) Hanley and Belfus, Inc., Philadelphia, PA, pp. 721-738 (1987).

3.   Thorne, P.S., Hillebrand, J., Magreni, C., Riley, E.J. and Karol, M.H. Experimental sensitization to subtilisin. I. Production of immediate- and late-onset pulmonary responses. Toxicol. Appl. Pharmacol. 86: 112-123 (1986).

4.   Ellakkani, M.A., Alarie, Y.C., Weyel, D.A., Mazumdar, S. and Karol, M.H. Pulmonary reactions to inhaled cotton dust: An animal model for byssinosis. Toxicol. Appl. Pharmacol. 74: 267-284 (1984).

5.   Ellakkani, M.A., Alarie, Y., Weyel, D. and Karol, M.H. Chronic pulmonary effects in guinea pigs from prolonged inhalation of cotton dust. Toxicol. Appl. Pharmacol. 88: 354-369 (1987).

6.   Griffiths-Johnson, D., Ryan, L. and Karol, M.H. Development of an animal model for organic dust toxic syndrome. Inhal. Toxicol. (1991) (in press).

7.   Alarie, Y., Kane, L. and Barrow, C. Sensory irritation: the use of an animal model to establish acceptable exposure to airborne chemical irritants. In: Toxicology: Principles and Practices, Vol. 1. (A.L. Reeves, Ed.) John Wiley and Sons, Inc., New York, NY, 1980, pp. 48-92.

8.   Hakkinen, P.J. and Witschi, H.P. Animal Models. In: Toxicology of Inhaled Materials. (H.P. Witschi and J.D. Brain, Eds.) Handbook of Experimental Pharmacology Vol. 75, Springer-Verlag, Berlin Heidelberg, 1985, pp. 95-114.

9.   Karol, M.H., Thorne, P.S. and Hillebrand, J.A. The immune response as a biological indicator of exposure. In: Occupational and Environmental Chemical Hazards (V. Foa, E.A. Emmett, M. Maroni, A. Colombi, Eds.) Ellis Horwood Ltd., Chichester, 1987, pp. 86-90.

10.   Karol, M.H., Dixon, C., Brady, M. and Alarie, Y. Immunologic sensitization and pulmonary hypersensitivity by repeated inhalation of aromatic isocyanates. Toxicol. Appl. Pharmacol. 53: 260-270 (1980).

11.   Weyel, D.A., Ellakkani, M., Alarie, Y. and Karol, M. An aerosol generator for the resuspension of cotton dust. Toxicol. Appl. Pharmacol. 76: 544-547 (1984).

12.   Karol, M.H., Hillebrand, J.A. and Thorne, P.S. Characteristics of weekly pulmonary hypersensitivity responses elicited in the guinea pig by inhalation of ovalbumin aerosols. Toxicol. Appl. Pharmacol. 100: 234-246 (1989).

13.   Thorne, P.S. and Karol, M.H. Association of fever with late-onset pulmonary hypersensitivity responses in the guinea pig. Toxicol. Appl. Pharmacol. 100: 247-258 (1989).

14.   Stadler, J. and Karol, M.H. Experimental delayed-onset pulmonary hypersensitivity: Identification of retest reactions in the lung. Toxicol. Appl. Pharmacol. 65: 323-328 (1982).

15.   Frigas, E. and Gleich, G.J. The eosinophil and the pathophysiology of asthma. J. Allergy Clin. Immunol. 77: 527-537 (1986).

16.   Jin, R. and Karol, M.H. Specificity of antibodies to toluene diisocyanate identified in workers and induced in an animal model. Am. Rev. Respir. Dis. 143: A439 (1991).

17.   Karol, M.H. Concentration-dependent immunologic response to toluene diisocyanate (TDI) following inhalation exposure. Toxicol. Appl. Pharmacol. 68: 229-241 (1983).

18.   Karol, M.H. Survey of industrial workers for antibodies to toluene diisocyanate. J. Occup. Med. 23: 741-747 (1981).

19.   Armour, C.L., Black, J.L. and Johnson, P.R.A. A role for inflammatory mediators in airway responsiveness. In: Mechanisms In Asthma: Pharmacology, Physiology and Management (C.L. Armour, J.L. Black, Eds.) Alan R. Liss, Inc., New York, NY, 1988, pp. 99-110.

20.   Thorne, P.S. and Karol, M.H. Assessment of airway reactivity in guinea pigs: Comparison of methods employing whole body plethysmography. Toxicology 52: 141-163 (1988).

21.   Popa, V., Douglas, J.S. and Bouhuys, A. Airway responses to histamine, acetylcholine, and propanolol in anaphylactic hypersensitivity in guinea pigs. J. Allergy Clin. Immunol. 51: 345-356 (1973).

22.   Griffiths-Johnson, D. and Karol, M.H. Validation of a non-invasive technique to assess development of airway hyperreactivity in an animal model of immunologic pulmonary hypersensitivity. Toxicology 65: 283-294 (1991).

23.   Rask-Andersen, A. Organic dust toxic syndrome among farmers. Brit. J. Ind. Med. 46: 233-238 (1989).

24.   Stober, W., Morrow, P.E. and Morawietz, G. Alveolar retention and clearance of insoluble particles in rats simulated by a new physiology-oriented compartmental kinetics model. Fund. Appl. Toxicol. 15: 329-349 (1990).

Antigen Specific
and Antigen Non-Specific Immunization

Robert Burrell

The word "hypersensitivity" in the multiple chemical hypersensitivity epithet means quite different things to those depending on whether they be toxicologists or immunologists. The term unfortunately prejudices immunologists because hypersensitivity has very precise connotations based on several decades of molecularly oriented studies which concluded that classical immunological reactions are the result of highly specific molecular interactions between precisely defined prosthetic groups on antigens (epitopes) reacting intimately with either specific immunoglobulins induced by prior contact with that antigen or with T lymphocyte antigen specific receptors (TCR). Although a certain degree of cross reactivity among different molecules can be expected, such may be explained on the basis of molecular structures having closely identical configurational fits. To suggest that chemically, quite-unrelated molecules cross react in the classical immunological sense is utterly foreign to those steeped in the lore of the enormous number of studies of such pioneers as Landsteiner, Heidleberger, Haurowitz, Campbell, and many other immunochemists. The studies of Landsteiner (1945) on the specificity of serological reactions spanned nearly four decades and one of his innumerable studies is summarized in Fig. 1. His method of diazotizing defined haptens on carrier molecules to obtain antisera and then reacting these antisera with homologous and closely related haptens forms one of the basic foundations of classical immunology.

Later, such reactions were placed on a more quantitative basis by people like Eisen (1990) who used the techniques of physical chemists and enzyme kineticists to study these same kinds of hapten-specific interactions. Through the use of equilibrium dialysis, one can obtain such information as antibody valence and degree of heterogeneity in an antibody population and even the affinity constant of an antigen-antibody reaction (Fig. 2). From these kinds of studies, it soon became evident that antigens and antibodies did not react by forming covalent bonds, but rather were held together by a number of physical forces. These forces (hydrogen bonding, electrostatic, Van der Waals, etc.) in themselves are non-specific and ordinarily are relatively weak, but become quite strong if the two molecules are very close together. Van der Waals forces for example decrease inversely with the seventh power of the distance separating two molecules. But what allows them to approximate themselves to each other so closely is the fine degree of complementary, struc-

GIVEN: An antiserum to **Meta-Aminobenzene** Sulfonic Acid

Reactive Haptens

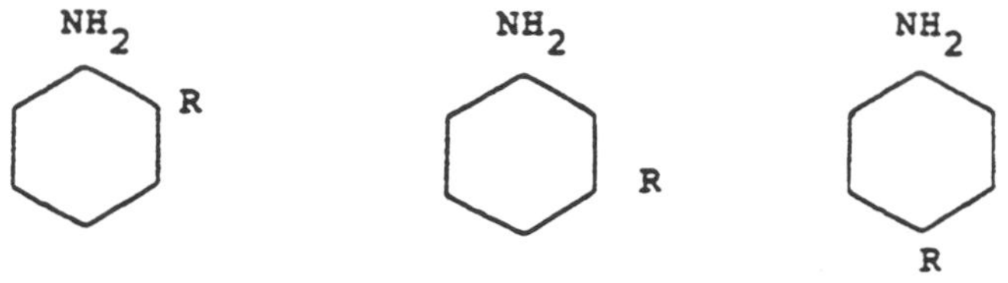

ORTHO-   META-   PARA-

Rx Antigen:

	ORTHO-	META-	PARA-
Homologous	+	+++	±
Aminobenzene As	-	+	-
Aminobenzoate	-	±	-

If R is large, cross reaction is even less because such a molecule can not make a close fit.

Fig. 1   One of the clasical experiments of Landsteiner illustrating the use of antibodies directed to synthetic haptens. In this case, antibodies were prepared against the hapten, meta-aminobenzene sulfonic acid conjugated to an appropriate carrier. Maximum serologic reactivity was obtained only when this antiserum was reacted with the hapten used in its production and not with molecules that were closely rleated chemically. Further, the specificity even extended to the position on the benzene ring on which the sulfonic group was attached.

tural configuration in the respective molecules. If a closely-related molecule has an additional conformational change, even moving a prosthetic group from a para- to a meta-position (Fig. 1), then that close fit is greatly compromised and there is either no reaction or a very much weaker one.

The availability of myeloma immunoglobulins, for which antigenic reactivity is known, afforded the immunochemist with a homogenous population of antibodies whose light and heavy chain variable domains could be sequenced. Now the exact molecular structure of the antibody as well as the antigen could be known. The amino acid sequence of a polypeptide determines how that peptide will fold three-dimensionally in space and that degree of folding is the antibody characteristic which allows the complementary fit of its homologous antigen. Fig. 3 shows the key amino acids in light and heavy chains that are in most intimate contact with a phosphorylcholine hapten. Essentially, the antigen fits in a very tight cleft formed by the folding of the light and heavy immunoglobulin chains. The lock-and-key analogy used for so many decades to explaining antigen-antibody fit has proven to be remarkably accurate.

Antibody Prepared Against	Test Hapten	"Average Affinity" K_{os}, liters mole^{-1} × 10^5
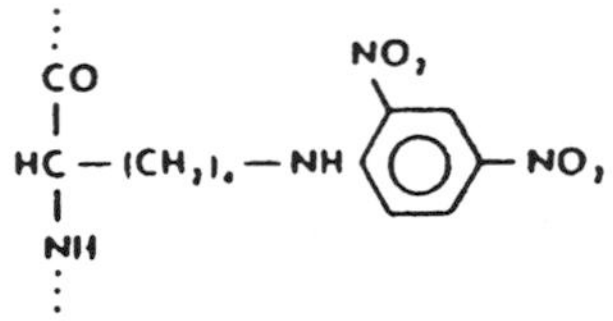 2,4-dinitrophenyl ι-lysyl group of Dnp protein	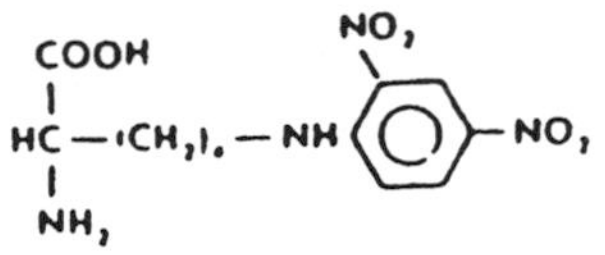 ε-Dnp-ι-lysine	200
	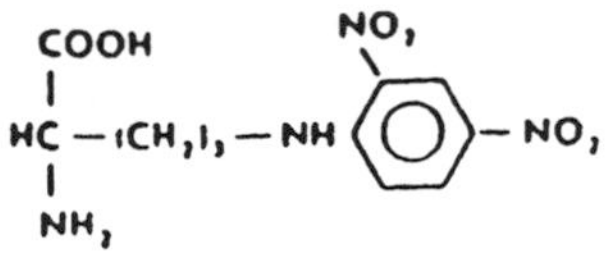 δ-Dnp-ι-ornithine	80
	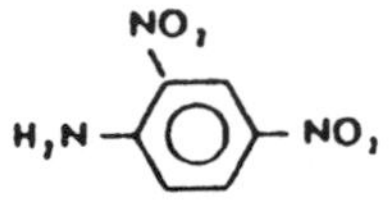 2,4 dinitroaniline	20
	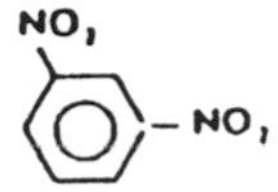 m-dinitrobenzene	8
	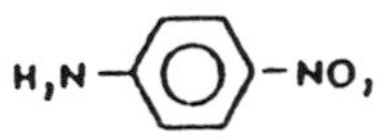 p-mononitroaniline	0.5

Fig. 2   When serological studies are extended quantitatively, affinity constants can be obtained using equilibrium dialysis methods. Note that the maximum affinity constant occurs only when the antiserum is reacted against the hapten used in its production and that this constant rapidly declines, the further removed the chemically substituted molecules are that are used in reacting against the same antiserum. From H.N. Eisen, General Immunology, 3rd Edition, 1990, Lippincott, Philadelphia, p. 17.

Contemporary molecular immunology is now focussed on the cellular aspects of the immune response, particularly the afferent induction pathways. The pathways involve first contacting the antigen, then processing the antigen so that peptide fragments of the native antigen are presented on the cell membrane surfaces of antigen presenting cells in context with major histocompatibility (MHC) molecules. This presentation is made to T helper lymphocytes who express receptors (TCR) for the presented antigen within this MHC context. Representative TCRs have also been sequenced and their structures can be considered analogous to antibody molecules (Fig. 4). Processed antigen molecules fit within the TCR cleft in a manner analogous with the way antigen fits in an imunoglobulin cleft. Thus even antigen-lymphocyte interactions are governed by the same kinds of intermolecular constraints that antigens and antibodies are subject to.

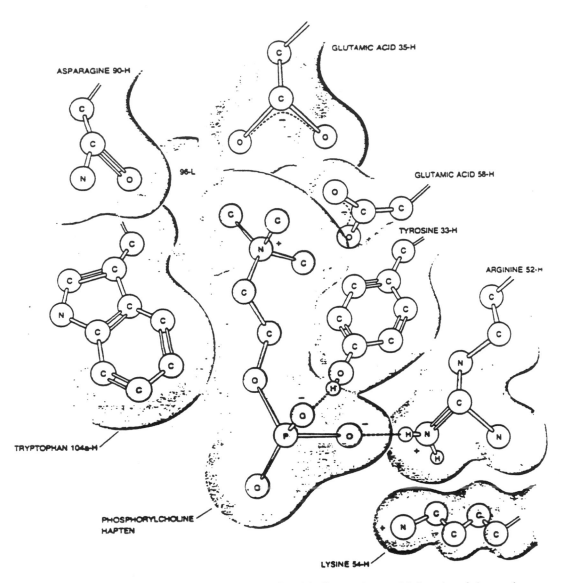

Fig. 3   Model of the binding of the hapten phosphorylcholine to the combining site of the myeloma
protein with which it is reactive. The numbers refer to the amino acid position in the heavy (H) or
light (L) chain respectively which is in closest approximation with the key polar forces on the
haptenic molecule. The way in which the peptide chains of the antibody fold provide for a
three-dimensional cleft in which the hapten may be attracted in a very close, conformational fit.
From Scientific American, 1977, 236(1):58.

But with the tremendous increase in modern knowledge of immunity has come the
realizations that 1) the immune system is far more complicated than intimate reactions
between closely-fitting, complementary molecules and 2) that the very same immune system

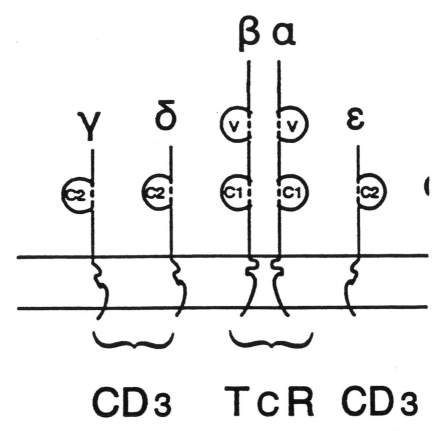

Fig. 4   A stylized model of the T cell receptor (TCR) molecule showing the resemblance to an immunglobulin molecule. It is composed of two chains embedded in the lymphocyte membrane. These chains also have variable and constant domains in a manner analogous to immunoglobulins. The CD3 chains are associated in proximity to the TCR and participate in stabilization of the molecular interactions with specific antigen and antigen- presentation markers.

that recognizes specific antigens also can be stimulated non-specifically to carry out its effects independently of antigens. Knowledge emanating from studies of such matters as the plethora of cytokines required for immunological reactions, the many forces involved in priming and translating immune recognition into inflammation, the pharmacology of immune reactions, the ability of some very important immunological cells to recognize for-eignness in an antigen-independent manner (e.g., NK cells), and the growing awareness of how the central nervous system and the immune system are inter-connected are a few examples that suggest that the immune system can play an important role in effecting antigen non-specific reactions as well.

Table 1 compares classical, antigen specific immunity with that of antigen non-specific reactions from several aspects. Non-antigen specific reactions are triggered by many kinds of agents, e.g., polysaccharides that activate the alternative complement pathway in the absence of antibody or tumor cells that activate NK cells. Many types of immuno-

TABLE 1

Comparison of Antigen Specific and Antigen Non-specific Immune Reaction

Feature	Antigen Specific	Antigen Non-specific
Stimulus	Antigen	Immunomodulator
Response measured	Antibodies T_c cells	Mediator production NK cells Pathology
Secondary	Anamnestic	Enhanced, Ag non-spedcific
Longevity	Long, Ag-dependent	Temporary

modulators or substances that mimic allergies can fall into this category. Although one usually measures humoral antibody or T cytotoxic cell responses from antigen-driven reactions, increases in mediator production or responsiveness of target tissues are the parameters usually assessed in non-antigen specific reactions. One of the earmarks of classical immune reactions is the anamnestic response, but increased reactivity can also be brought about by a number of non-antigen specific mechanisms as well; the Shwartzman reaction is a good example. Longevity of antigen specific reactions is mostly dependent on dose and presence of antigen as well as the induction of memory cells. Non-antigen specific reactions are often temporary, that is enhanced responsiveness is only present within a short time range at such time when two independent stimuli have been received by the target tissue. At other times, increased hyperreactivity may be more permanent as a result of a major event such as an infection which then renders the subject hyperresponsive at other times years later, especially when another infection or irritation may take place. Table 2 lists several examples of antigen non-specific reactions.

Classical immunology of course is concerned with the effects of antigen on antigen-presenting cells, all classes of lymphocytes, antigen processing, effector regulation and the like, but these same processes are either closely allied or one and the same as the inflammatory mechanisms generated by the induction of immune responses. Even for antigen specific responses, it has been pointed out by Warren and Chedid (1987) that if immunology is defined as the body's reaction to foreign insult, then one must also consider such non-antigen specific, induced responses as the involvement of endothelial cells, activation of the complement and procoagulation cascades, vascular changes, cytokine generation, lysosomal enzyme release, the generation of toxic oxygen radicals, arachidonic acid metabolism, even the effect of interleukin 1 (IL-1) on the hypothalamus, among many others (Table 3).

The most important factor in trying to explain any hypersensitivity reaction is the matter of dealing with why, of all of those exposed to a given agent, only a small subset respond in such a deleterious manner. Classical allergists at least have an explanation largely in the form of antigen-specific IgE or lymphocytes that have been produced.

Explaining individual idiosyncrasies that are not allergy-based is difficult owing to the

TABLE 2

Examples of Antigen Non-Specific Reactions

Activation of alternative C pathway
    Peptigoglycans
    Endotoxin
    Thrombin

NK cell recognition
    Tumor cells
    Virus-infected cells?

Acute phase reactants
    C-reactive proteins
      Interaction with bacterial polysaccharides
    Haptaglobin
      Anti-inflammatory factor
    Low density lipoprotein
      Partial detoxification of bacterial endotoxin

dearth of studies on other possible explanations. Pharmacologists recognize idiosyncratic reactions to drugs and account for them by a number of physiologic or pharmacologic host factors that determine reactivity to such agents. Those who manifest hyperresponsiveness to low dose levels have a genetic component that is responsible for either how the drug is metabolized or how that person's cellular/molecular mechanisms respond. These pharmacogenetic bases are further complicated by the numerous types of environmental factors that affect gene expression or the occurrence of the clinical manifestation. An individual with an abnormal gene product would never be noticed unless that person becomes exposed to an agent in which there is an interaction with that product. In these instances, only a certain subset of the population is going to metabolize the agent in such a way as to result in toxic by-products. When an environmental stimulus is additionally required to express the abnormal gene action, as in the case of cutaneous photosensitization in certain porphyrias, clinical manifestations might be erroneously considered immune in origin, but the interesting thing here is the requirement for two separate signals. In addition to genetic factors, other conditions which are known to alter pharmacologic reactivity to substances are age, presence of concurrent disease, nutritional status, hormone levels, diurnal variation in drug disposition or sensitivity, environmental factors that induce or inhibit drug metabolism, and efficiency of repair mechanisms (Nebert and Weber, 1990).

    Although a fair amount is known about pharmacogenetics, much less is know about genetic factors that affect physiologic responsiveness. Some of the host factors that definitely have bases contributing to the way individuals hyperrespond to environmental agents are airway hyperresponsiveness to methyl choline challenge, increased propensity of basophils to release histamine, dermatographism, and hyperreactivity to histamine.

    In MCS, not only are we faced with the problem of explaining hyperresponsiveness, we

TABLE 3

Examples of Host Responses that Must Also Be Considered Immunological

Endothelial cell reactions

Activation of C cascades in the absence of antibody

Activation of procoagulant activity

Acute phase reactant responses

Aracidonic acid metabolism

NK cell enhancement

Lysosomal enzyme release

Generation of toxic oxygen radicals

IL-1 effect on the hypothalamus

Beta-endorphin activity

Generation of neuropeptides

have an additional problem of explaining how this hyperresponsiveness is directed towards numerous agents with no obvious molecular relationships in a given individual. One of the environmental factors that has been studied more concerning the subsequent development of hyperresponsiveness has been that of association of certain types of infections with triggering development of bronchial asthma particularly in children and to some extent in adults. But even in these instances, other factors are additionally required, e.g., the type of virus causing the infection, the age at which the infection first takes place, the presence of certain constitutional symptoms at the time of infection, genetic predisposition, pre-existing airway hyperreactivity, and sex. Busse (1990) has recently reviewed the possible mechanisms (summarized in Table 4) by which this altered reactivity might occur and among them proposed are that certain viruses, notably respiratory syncitial virus, may enhance IgE synthesis and mediator release from basophils either directly or through the action of interferon from virus-stimulated lymphocytes. The viruses may also interfere with beta-adrenergic function either by directly inhibiting such function, by passively allowing the mast cell to promote more mediator release, or by enhancing cholinergic stimulation of airways smooth muscle. Finally, the virus may desquamate epithelium thus exposing afferent neurons to stimulation and decreasing the amount of an endopeptidase that normally degrades the amount of the peptidergic smooth muscle contractant, substance P. Thus, extrinsic asthma may require 1) previously altered tissue growth/damage from an infectious agent or an irritant, 2) genetic predisposition, and 3) a subsequent trigger later in

TABLE 4

Possible Mechanisms to Explain Asthma Predisposing Factors of Respiratory Infections

1. Viruses induce polyclonic IgE antibody responses
   Genetic predisposition
   Associated with particular viruses, e.g. RSV
   Elevated airway histamine
   Histamine levels correlate with hypoxia

2. Viruses enhance basophil mediator release/chemotaxis
   Can substitute interferon for virus

3. Viruses may diminish beta-adrenergic function
   Could establish a permissive situation promoting mast cell mediator release

4. Viruses promote cholinergic-dependent responses
   Not in smooth muscle directly, but in inflammatory cell mediator release

5. Viruses affect peptidergic responses
   Damage airway epithelium
   Affect bronchodilator effects of LPO products
   Unopposed constriction of Substance P
   Infection inhibits SP degrading enzyme

HYPOTHETICAL SCENARIO

   Inflammatory granulocytes, altered by infection
   Recruited to the airways
   Enhanced mediator release *in situ*
   Leads to airway hyperresponsiveness
   Must take place in the generally-predisposed

life. We might ask with regard to multiple chemical sensitivity, what the role might be of a previous infection or toxic event at a critical point of development in a genetically-predisposed individual such that neurologically controlled effector mechanisms become subsequently more hyperreactive. But again, in trying to explain MCS, one confronts the observations that its signs and symptoms are not referrable to specific shock organs such as bronchial smooth muscle in asthma. Instead, many of the symptoms are so vague, subjective, and rarely subject to measurement.

Another reason certain individuals might be excessively responsive at a given time might be because primary, incidental stimuli may serve to prime inflammatory mediators that temporarily render that individual more responsive to a second, unrelated stimulus. The coincidental exposure to two or more stimuli within a given time may induce synergistically-altered responses whereas if these same two unrelated stimuli were received at separate times further apart, there would be no abnormal reaction. Much of what we know about this kind of exaggerated responsiveness has been learned from the study of host response to bacterial endotoxin and the prototype of this type of response is the Shwartzman reaction. The endotoxin-induced Shwartzman reaction can be manifest either in a dermal or systemic form--both lead to increased target organ hyperreactivity following exposure to the second stimulus and thus meets the definition of altered reactivity. Neither is antigen specific reaction because the two doses do not need to be antigenically related, the time between sensitization and provocation is too short to induce antigen specific reactivity, and the increased sensitivity does not last more than a few days.

TABLE 5
___

Examples of "Second Signal" Mediator Responses
___

Shwartzman reaction
Adult respiratory distress syndrome
Endotoxin enhanced responsiveness
Infection-asthma
Receptors-cAMP/arachidonic acid metabolites
Immunomodulator/adjuvants
Interferon stimulation
Pharmacogenetic hyperresponsiveness

It is now felt that this type of temporal, increased reactivity may be due to several mediator systems being turned on to such an extent by various types of priming stimuli that occur prior to exposure to a further stimulant. Much attention has been directed to the idea of endotoxin-initiated "second signals" wherein endotoxin primes target cells for subsequent enhanced reactivity. This is an exceedingly complex area and could involve several different types of systems including calcium ion channel triggering, activation of cyclic nucleotide synthesis, increased toxic oxygen radical formation, and enhanced cytokine production, e.g., TNF and IFN- (Heremans et al., 1990), among others (see Table 5 for examples). This may relate to multiple chemical sensitivity in that if an individual encounters an immunomodulating or pharmacologically-altering agent at near the same time he encounters a triggering substance, the combined effects of mediators from the two stimuli may synergistically cause an exaggerated reaction.

REFERENCES

Busse, W.W. 1990. Respiratory infections: Their role in airway responsiveness and the pathogenesis of diseases and asthma. J. Allergy Clin. Immunol. 85:671-683.

Eisen, H.N. 1990. General Immunology, pp. 11-18. Philadelphia:J.B. Lippincott.

Heremans, H., J. Vandamme, C. Dillen, R. Dijkmans, and A. Billiau 1990. Interferon-gamma, a mediator of lethal lipopolysaccharide- induced Shwartzman-like shock reactions in mice. J. Exp. Med. 171:1853-1869.

Landsteiner, K. 1945. The Specificity of Serological Reactions. 450 pp. Boston:Harvard University Press.

Nebert, D.W., and W.W. Weber 1990. Pharmacogenetics. In Principles of Drug Action, eds.

W.B. Pratt, and P. Taylor, pp. 469-531. New York:Churchill Livingstone.

Warren, H.S. and Chedid, L.A. 1987. Strategies for the treatment of endotoxemia: Significance of the acute-phase response. Rev. Infect. Dis., 9 (Supp 5), S630-S638.

Neuropsychiatric & Biopsychosocial Mechanisms In Multiple Chemical Sensitivity: An Olfactory-Limbic System Model

Iris R. Bell

INTRODUCTION

The purpose of this paper is to review the clinical and research psychiatric and psychophysiologic literature on multiple chemical sensitivity patients (MCS) and to develop an integrative neuropsychiatric and biopsychosocial systems model of possible mechanisms. Much of the controversy over MCS has focused on hypotheses that the clinical syndromes must derive exclusively either from physiological or from psychogenic sources. However, such a dichotomous view of MCS is overly simplistic. A large body of data from both human and animal studies supports a more complex biopsychosocial approach involving an interplay of multiple influences in the expression of all human illnesses (Engel, 1977; Schwartz, 1982), including the clinical phenomenology of chemical sensitivities (Bell, 1987). This model means that MCS would involve a continuum and interaction of mechanisms in terms of the relative contributions of biological, psychological, and social factors in a given patient. Genetically-based neurochemical and/or receptor vulnerabilities in the central and autonomic nervous systems would make certain subsets of the population more likely to experience adverse effects of low dose chemical exposures.

Furthermore, even psychological and social "stress" can express health effects via biological mechanisms (Shavit et al., 1984; Sapolsky et al., 1990). Most of the major psychiatric disorders have genetic and neurochemical components (e.g. monozygotic twins have a 65-75% concordance vs dizygotic twins with a 14-19% concordance for major depression - Talbott et al., 1988). Thus, evidence for one type of factor (e.g. psychological) cannot rule out the contribution of any other type of factor (e.g. biological) to a clinical syndrome. Appropriate research in MCS must test plausible hypotheses by which a particular factor could causally contribute to or ultimately produce the clinical picture. Phenomenological labels alone fail to meet the latter requirement.

A logical site to examine the potential convergence of biological, psychological, and social factors in MCS is the central nervous system (CNS), especially the interconnected olfactory and limbic systems (Bell, 1982; Ashford and Miller, 1991). The CNS receives input from and sends output to other subsystems of the body, including behavioral, endocrine, immune, and autonomic functions, each of which may be disturbed in certain MCS patients.

The CNS is an integrative center that transduces biological and psychosocial experiences into changes in neural activity; in turn, these send biological, psychological, and social behaviors back out into the environment. Many pesticides and solvents, which are putative triggers for MCS, have effects on central nervous system function (Isaacson et al., 1990; Llorens et al, 1990; Tham et al., 1990) that could interact with individual vulnerabilities.

The olfactory system is the usual pathway by which airborne chemicals from the external environment interact with the brain. The limbic system is the phylogenetically older part of the CNS that receives extensive input from the olfactory system and regulates numerous functions including mood and social behaviors, cognitive function, and eating/drinking/ reproductive behaviors (Cain, 1974). Dysfunctions of the limbic system may play a key role in unipolar and bipolar depression, panic and other anxiety disorders, and schizophrenia. Damage to the hippocampus in the limbic system is an important feature of dementia, especially of the Alzheimer type (Sapolsky, 1990).

CONTROLLED STUDIES ON
MULTIPLE CHEMICAL SENSITIVITY PATIENTS

At the outset, a model for MCS must be able to account for the clinical phenomenology. Patients with MCS report multiple symptoms in multiple systems. Clinical observations also include initial sensitization with acute high dose exposure, chronic low-dose sensitivity, spreading of number of sensitivities from one to many chemicals, adaptation to chronic exposures, and concomitant multiple food sensitivities (MFS). Little rigorous data are available to characterize the MCS population and their responses to chemical challenges (Tables 1a and 1b).

TABLE 1a.

Multiple Chemical and Food Sensitivity Studies: Patient Differences from Controls

Study	Controls (n)	Patients (n)	Current or past Depression	Cognitive Problems
Doty (1988)	normals (18)	MCS (18)	yes	Problems
Staudenmayer (1990)	normals (55)	MCS (58)	N/A	Not tested
Simon (1990)	chem-exposed normals (23)	MCS (13)	Yes	Not tested
Black (1990)	normals (33)	MCS (26)	Yes	Not tested
Bell (1990)	normals (23)	MFS with MCS (25)	Yes	Yes

TABLE 1b.

Multiple Chemical and Food Sensitivity Studies: Patient Differences from Controls

Study	Controls (n)	Patients (n)	ANS Sx	Other
Doty (1988)	normals (18)	MCS (18)	↑ resp. rate ↓ nasal patency = HR, = BP	N/A
Staudenmayer (1990)	normals (55)	MCS (58)	= skin temp = skin resist	↑ scalp EMG ↑ spectral EEG beta activity
Simon (1990)	normals (23)	MCS (13)	N/A	↑ SCL-90 somatization subscale ↓ Barsky Amplification Scale
Black (1990)	normals (33)	MCS (26)	N/A	↑ anxiety disorders ↑ somatization disorder
Bell (1990)	normals (23)	MFS with MCS (25)	↓ skin temperature in poor cognitive performers	+ tinnitus + daytime sleepiness

Despite extensive debate, only one controlled study has examined the effects of inhaled chemicals on individuals who report the MCS syndrome. Doty et al. (1988) compared the olfactory thresholds and selected autonomic responses to inhalation of phenyl ethyl alcohol and methyl ethyl ketone of 18 MCS patients with a matched control group of 18 normals. They found that the MCS group did not differ from normals on olfactory thresholds or cardiovascular responses to the two test chemicals, but that the MCS patients showed increased respiration rate and decreased nasal airway patency during chemical challenges. This study presented the chemicals in an environmentally controlled test chamber, but did not mask the identity of the chemical exposures or manipulate possible adaptation to the test items. Thus, these data confirm the presence of autonomic differences in MCS patients in terms of respiration and nasal airway function to specific chemical challenges, but do not address the larger clinical syndrome under consideration.

The remainder of the controlled data describes nonspecific phenomenology in subsets of

MCS patients. Most of these studies did not document tests of whether or not chemicals trigger illness in MCS patients. For example, Simon et al (1990) compared cohorts of plastics workers who did and did not develop chronic disabling MCS from low dose exposures to phenol, formaldehyde, and methyl ethyl ketone. They found that 54% of the 13 workers who developed MCS had a prior history of anxiety or depressive disorders in comparison with 4% of control workers. Staudenmayer and Selner (1990) observed higher levels of scalp electromyographic activity in MCS patients and similar patterns of electroencephalographic beta activity in MCS and mixed psychological outpatients in comparison with normal controls; this group reported performing negative controlled chemical challenges on the MCS subjects, but did not present their specific dependent measures or objective findings on this key point. Black's group (Black et al., 1990) reported that 65% of 23 MCS patients had histories of current or past mood, anxiety, or somatoform disorder in comparison with 28% of matched community controls. Although Black suggested that the psychiatric diagnoses could "explain some or all of their symptoms," they did not challenge these patients with chemicals to rule out chemically-induced illness. By consensual agreement within the American Psychiatric Association (Diagnostic and Statistical Manual III-revised, 1987), psychiatric diagnoses are descriptive labels only for phenomenology, not etiological or mechanistic explanations for syndromes. Thus, a psychiatric diagnosis labels a pattern of signs and symptoms, but offers no hypothesis concerning the mechanism(s) of the clinical phenomena (Davidoff et al., 1991). Finally, studies that have found psychiatric diagnoses in MCS patients have used non-random samples of the potentially chemically-sensitive population. Subject selection of MCS patients for research (Simon et al., 1990) and clinical series (Terr, 1986) has often derived from disabled subsamples with worker's compensation and other litigation claims, those self-identified with the field of clinical ecology (Doty et al., 1988; Black et al., 1990), or even those seen in psychiatric clinics (Black et al., 1990). Even in more recognized clinical syndromes involving chronic pain or traumatic injury, those individuals with associated legal and monetary cases represent a unique subset of any given chronically ill population. The discovery of psychiatric problems in a sample drawn in part from those seen in a psychiatric clinic (Black et al., 1990) is clearly a skewed representation of the true prevalence of psychiatric problems in the overall MCS population. Proper epidemiological research on characteristics of MCS will require case identification with more representative sampling from the general, chemically-exposed population.

Moreover, none of the studies that found increased depression, anxiety, and/or somatoform disorder histories compared MCS patients with chronically-ill control groups as opposed to healthy normals. Since chronic illnesses of all types, including cancer, heart disease, and autoimmune disorders, have a current or past prevalence of depression and anxiety disorders ranging from 5% to as high as 45% (general average, 20%) (Katon and Sullivan, 1990), the finding of depression and anxiety would be expected in chronically ill MCS patients and would not establish a specific causal link between the affective disturbance and the chronic illness. In addition, somatic concerns, which constitute clinical criteria for depression and somatoform disorders, are often excluded in research studies of medically ill populations in order to avoid false positive diagnoses of depression on largely somatic symptomatology. In contrast with other investigations on geriatric and medically-ill populations - which use modified interview and/or self-rating scales free of somatic bias or adjust statistically for medically-related somatic complaints, no study of the psychiatric syndromes associated with subsets of MCS patients has ever made any such methodological adjustments.

OLFACTORY-LIMBIC SYSTEM MODEL
AND AFFECTIVE SPECTRUM DISORDERS

Affective Spectrum Disorders

Ironically, the finding of psychiatric dysfunctions may nonetheless offer clues to the biology of the disorder in a subset of MCS cases. That is, the mechanisms of depression and anxiety, as well as other psychiatric symptoms, involve putative dysregulation of brain chemistry and neurotransmitter receptors in specific neural pathways (Reiman et al., 1986; Talbott et al., 1988). Thus, one might expect individuals vulnerable to major psychiatric disorders be among the most susceptible to low doses of those environmental chemicals that could worsen their inherent dysfunctions in brain chemistry, either by direct action or by activation of endogenous mediators (Bell, 1982).

For example, animal models of depression include drug-induced elevation of brain acetylcholine, drug-induced depletion of brain serotonin and/or norepinephrine, and destruction of the olfactory bulbs (Jesberger and Richardson, 1985, 1988; van Riezen and Leonard, 1990). Psychiatric investigators have already raised the possibility that the presence of major depression in certain patients indicates an increased neurochemical vulnerability to certain environmental chemicals such as organophosphate pesticides (Rosenthal and Cameron, 1991). The latter agents raise brain acetylcholine levels, an effect which can induce depression in susceptible persons (Dilsaver, 1986).

The overlap between the biology of depression and certain MCS and MFS syndromes is further supported by converging evidence that many of the same disorders that some clinicians have claimed to be triggered by adverse food and chemical reactions also improve during chronic treatment with antidepressant medications. The disorders within medicine that share responsivity to these drugs include major depression, panic disorder, bulimia, obsessive-compulsive disorder, attention-deficit disorder with hyperactivity, migraine headache, irritable bowel syndrome, and cataplexy (Hudson and Pope, 1990). Notably, 87% of patients with the somatization disorder known as Briquet's syndrome (hysteria) also experience major depression during the course of their illness; and 76% of such patients report multiple food intolerances as a criterion for diagnosis (Purtell et al., 1951), another overlap with MCS/MFS phenomenology. In addition, Bell et al. (1990) have recorded tinnitus in 48% of one sample of community-recruited MFS/MCS patients, a condition also known to be responsive to antidepressant medications (Sullivan, et al., 1989). In terms of environmental reactivity, it is striking that a number of studies have found persons with a history of major depression to have an inordinately high prevalence of atopic allergy histories in comparison with both normals and other types of psychiatric patients (Nasr et al., 1981; Sugerman et al., 1982; Bartko and Kasper, 1989; Marshall, 1989; Bell et al., 1991). Such data contrast with the current lack of evidence for atopic mechanisms in MCS or MFS, despite the depression histories. More thorough epidemiological investigation of this point is needed to clarify the prevalence of atopy in depressed versus nondepressed MCS/MFS patients. Interestingly, however, in addition to other mechanisms, several tricyclic antidepressants exert also significant antihistamine effects at both H1 and H2 receptors (Schatzberg and Cole, 1991).

From the above list, controlled studies have indicated that adverse food and ingested chemical reactions may worsen certain cases of ADD with hyperactivity (Egger et al., 1985; Kaplan et al., 1990; Swanson and Kinsbourne, 1980), migraine headache (Egger et al., 1983; Egger et al, 1989), and irritable bowel (Jones, et al., 1982), though some researchers have had negative findings in the latter disorder (Bentley et al., 1983). Thus, it is possible that some brain neurochemical dysfunction susceptible to worsening by adverse food and chemical reactions and to improving with antidepressant drugs is common across a range of disorders, not all purely "psychogenic" in nature. This hypothesis raises two possibilities in identification of possible MCS cases: (a) choosing subjects with personal and family histories loaded for more than one of the above disorders may optimize chances of finding true cases; for example, Egger's group observed that the MFS dietary responders were those with histories of both migraine and seizures, but not with seizures alone (Egger, et al., 1989); (b) screening subjects for those who improve from 4-6 week trials with antidepressant medications, regardless of clinical diagnosis, should assist selection of biologically more homogeneous patients.

Olfactory Bulbectomy Model of Depression

It is also striking that one of the best animal models of depression currently used by drug companies to identify new antidepressant medications is that of olfactory bulbectomy (Jesberger and Richardson, 1988; van Riezen and Leonard, 1990). As in human depression, animals with olfactory bulb damage demonstrate neuroanatomic dysregulation of the limbic-hypothalamic axis, neurochemical imbalances of serotonin, norepinephrine, gamma-amino-butyric acid, and acetylcholine, and improvement from chronic but not acute treatment with antidepressant drugs (Jesberger and Richardson, 1988). Furthermore, olfactory bulbectomy increases the vulnerability of another limbic structure, the amygdala, to kindling (see below). The amygdala is especially important in regulation of affect (e.g. fear-avoidance; rage), drive, and related autonomic/endocrine/immune functions (Mesulam, 1985). Given the role of the olfactory bulb in transmitting sensory and nonsensory information concerning airborne chemicals to the rest of the brain (Cain, 1974), it is reasonable to hypothesize that individuals with dysfunctional olfactory bulb pathways secondary to inherent neurochemical and/or receptor alterations may be the population most sensitive to developing MCS.

Anxiety Disorders and Environmental Chemicals

Persons prone to anxiety may also have inherent neurochemical defects that make them particularly susceptible to certain environmental chemicals. For example, drugs and certain pesticides such as lindane that may antagonize receptors for the inhibitory neurotransmitter gamma-aminobutyric acid (GABA) have apparent anxiogenic properties in animals and humans (Llorens et al., 1990), opposite to those of antianxiety drugs such as the benzodiazepines. Thus, persons with benzodiazepine-responsive anxiety disorders that place them closer to threshold for anxiety due to relatively impaired GABAergic function at baseline might be expected to experience increased anxiety on a biological basis from lindane or related exposure. Another affective spectrum anxiety disorder that a small subset of MCS patients may experience is panic disorder (Dager et al., 1987). Panic disorder is a chronic disorder associated with a broad range of symptoms, especially recurrent acute attacks of difficulty breathing and hyperventilation, palpitations, dizziness, fatigue, and an impending sense of doom. Chemical agents such as low concentrations of carbon dioxide (e.g. 5.5%) in room air and as excess lactate trigger panic attacks in panic patients much more often than in normals under open and blind conditions (Woods, et al., 1986; Roy-Byrne, et al., 1988). In view of the olfactory-limbic model for MCS, it is notable that panic disorder patients have hemispheric asymmetry of blood flow in parahippocampal brain regions on positron emission tomography scans (Reiman et al, 1986). The parahippocampus is a limbic site that receives input from the main olfactory bulb, a major way-station for input from the nose and olfactory nerves. Furthermore, Dager et al. (1987) have hypothesized that occupational solvent exposures could induce panic disorder, possibly by activating a kindling process in the limbic system. Thus, a plausible hypothesis for some cases of environmentally-activated panic disorder is a biologically-based dysfunction of limbic pathways in response to olfactory system input, triggered by certain environmental chemicals.

Kindling and Long-Term Potentiation in the Limbic System:
Possible Role in the Mechanisms of MCS

Kindling occurs when repeated subthreshold stimuli summate to trigger seizure activity in brain cells with previously normal activity. The limbic system is especially susceptible to kindling, which may also play a role in the developing chronicity and treatment resistance of certain bipolar affective disorders (Post et al., 1984). In view of the hypothesized responsiveness of affective spectrum disorders to antidepressants, it is notable that these drugs reportedly raise the threshold for or even prevent kindling (cf. Jesberger and Richardson, 1985). Therefore, another, more mechanism-oriented label for the spectrum of disorders that is common to both affective spectrum and MFS/MCS might be kindling-related syndromes potentially responsive to antikindling medications.

Another property of limbic brain cells perhaps related to kindling is that of long-term potentiation (LTP). LTP involves persisting enhancement of synaptic response initiated by brief high-frequency stimulation of excitatory pathways at subictal levels (Racine et al., 1983; Stripling et al., 1988; Jung et al., 1990; Kanter and Haberly, 1990). LTP is hypothesized to subserve memory formation and retrieval in the hippocampal region of the limbic system, but also occurs in olfactory cortex following high-frequency stimulation of the olfactory bulb (Stripling et al., 1988).

Notably, the nature of kindling - and perhaps long-term potentiation - suggests that these biological mechanisms might participate in development and/or expression of MCS - that is, by either initial acute high doses (cf. high doses=high frequency stimuli) or chronic summated low doses (cf. low doses=low frequency stimuli) of environmental chemicals stimulating the firing, long-term potentiation, and/or kindling of olfactory pathways - and from there, many other regions of the limbic system in genetically-predisposed individuals. Indeed, Bokina et al. (1976) reported a rabbit model of low dose environmental pollutant intolerance that matches the latter hypothesis. Their group noted that low concentrations of pollutants combined with a functional load (rhythmic flashing light) triggered abnormal paroxysmal activity in the olfactory bulb and corticomedical amygdala after an initial exposure to high concentrations of the same pollutants. Furthermore, Bokina reported alterations in visual evoked potentials during chronic exposure to low concentrations of formaldehyde and ozone.

In summary, olfactory pathway kindling and LTP might participate in registering information concerning past high dose and/or cumulative low dose chemical exposures that would increase the likelihood of limbic responsivity to subsequent low dose exposures. Resultant dysfunctions in numerous behavioral, autonomic, and endocrine subsystems under limbic regulation might then lead to multiple symptoms through a range of end-organ mechanisms. Finally, it is also possible that chemical exposures which kindle certain limbic pathways could damage capacity for long-term potentiation in the hippocampus (Racine et al., 1983; Armstrong et al., 1990). Such events might in turn disrupt capacity for normal memory formation. These hypotheses offer the possibility of developing animal models for MCS at the level of the central nervous system that would permit greater experimental control and eliminate concerns regarding subject expectation and bias.

Brain Electrical Activity and Chemical Exposures

Although we have no further EEG data on MCS patients or chemically-exposed animals, Lorig and colleagues (Lorig and Schwartz, 1988; Lorig et al., 1988; Lorig, 1989; Lorig et al., 1990a; Lorig et al., 1990b) have performed a series of studies on the effects of low concentration odors on electroencephalographic responses, mood, and cognitive activity in normal human subjects (Lorig et al., 1990a; 1990b). Their findings suggest that chemical odors below the olfactory threshold (i.e. that are consciously undetectable) nonetheless produce distinct EEG responses, worsened mood, and poorer performance on a visual search task (Lorig et al., 1990b). They also examined evoked potentials during a common test of auditory attention and observed increased midline P200 wave amplitude even at an undetected low concentration of a common perfume constituent (galaxolide)(Lorig et al., 1990b). The P200 findings indicated that even undetected airborne chemicals may have a disruptive influence on attention processes. Moreover, in another investigation, they reported that EEG alpha frequency activity in the left hemisphere was lower and EEG beta activity showed greater spatial diversity between posterior and anterior regions during nose breathing than during mouth breathing of unfiltered indoor room air (Lorig et al., 1988). They concluded that undetected odors in indoor air inhaled via the nose exert a distinct effect on brain function outside conscious awareness or changes due to sensory perception (Lorig et al., 1988). Therefore, despite the subcortical location of the olfactory system, its broad connections into the forebrain provide the neuroanatomical and neurophysiological substrate for broad effects of low levels of environmental chemicals on EEG and behaviors regulated by the limbic system.

If such effects are measurable in normals, the olfactory-limbic system model for MCS would predict that the EEG, evoked potential, and cognitive findings would be even more pronounced in MCS patients. Preliminary evidence does support the possibility of abnormalities of attention and concentration in MCS. For example, Bell et al. (1990; submitted for publication, 1991) noted significantly slower performance on a timed mental arithmetic task in outpatients with multiple food and chemical sensitivities than in normal controls. Over 70% of this same sample of patients reported the symptom of difficulty concentrating in comparison with 13% of the controls (p < 0.008). In addition, Feidler and Kippen (this volume) have reported evidence of poorer overall performance by MCS patients on the California Verbal Learning Test, a measure of immediate, delayed, and recognition memory also sensitive to attention processes.

In brief, the limbic-olfactory model suggests that even unperceived low concentrations of airborne chemicals can induce changes in brain electrical activity and resultant behaviors and physiology in both animals and human subjects. In certain individuals with particular neurophysiological and/or neurochemical predisposition to limbic system dysfunction (e.g. depression or anxiety disorders), such chemical exposures would activate clinical syndromes described as MCS. The environmental chemicals enter the brain via the olfactory system (Cain, 1974; Shipley, 1985) and would mobilize the abnormal brain processes in the limbic brain with or without sensory awareness via kindling and/or long-term potentiation with initial high dose exposures or with repeated low concentration exposures. One important marker of exogenous chemical olfactory-limbic effects may be abnormalities of attention processes, which could be quantified with neuropsychological and EEG evoked potential tests during undetected chemical as well as sham exposures. Animal models may permit depth electrode recordings with behavioral learning tests, while MCS patients may provide surface computerized EEG and evoked potential data with attention and memory tests to evaluate the above hypotheses (cf. James et al., 1987). Time factors in terms of sequencing high to low dose exposures and of giving repeated low dose exposures may be necessary to elicit the proposed phenomena.

PSYCHOSOCIAL MODULATORS OF BIOLOGICAL EVENTS

Within its descriptive labels, modern psychiatry has embraced a biopsychosocial perspective for diagnosis. The DSM-IIIR (American Psychiatric Association, 1987), for example, employs a multiaxial system to maximize appreciation of the multidimensionality of psychiatric disorders. Thus, axis I covers clinical syndromes, axis II personality disorders, axis III medical disorders, axis IV severity of psychosocial stressors, and axis V global assessment of functioning. Obviously, patients with psychotic delusions of chemical sensitivity, those with only personality disorders, or those with completely incorrect, though nonpsychotic, misattribution of symptoms to chemical reactions may exist within the continuum of patients claiming to have MCS. At least in the case of psychosis, however, none of the studies that assessed psychiatric status reported finding any psychotic type disorders in certain MCS samples (Doty et al., 1988; Black et al., 1990; Simon et al., 1990; Staudenmayer and Selner, 1990). More refined and objective measures for identification of true cases will assist in screening out such individuals as well as any cases of nondelusional misattribution from MCS study populations.

Nevertheless, illnesses generally have psychosocial modulators, which interact with underlying biological vulnerabilities. These modulators include: premorbid personality and coping style (McCrae and Costa, 1986; Jamner et al., 1988); perceived social support (Cohen and Syme, 1985); perceived familiarity of the surrounding environment (Siegel et al., 1982); sense of control over the illness (Rodin, 1986; Sanderson et al., 1989); and classical conditioning of responses to both triggering and treating agents (Eikelboom and Stewart, 1982; Bolla-Wilson, et al., 1988; MacQueen et al., 1989). As emphasized above, identification of these factors in patients with MCS would be expected. For example, MCS patients have organized local and national support groups, which may have both positive benefits in terms of providing buffering effects from social support networks and negative effects in terms of reinforcing illness behaviors (Bell, 1987). They merit specific study to optimize clinical care and to understand the MCS syndromes, but demonstration of these psychosocial phenomena would not by itself "disprove" a role for biological causes of the syndromes.

At the same time, coincidental or deliberate manipulation of these factors would be likely to exert a marked influence on course and outcome in various individuals. Even in the case of terminal cancer, for example, a defiant and demanding versus submissive personality style (Levy, 1985) and participation in a cancer patient support group therapy versus no group therapy (Spiegel et al., 1989) are associated with longer survival times. Providing nursing home residents with a sense of control over their immediate physical surroundings and daily activities substantially reduces subsequent mortality rates over those of residents without such sense of control (Rodin, 1986). In the case of affective spectrum panic disorder, providing patients with an illusion of control over exposure to carbon dioxide will attenuate the severity of the carbon dioxide-induced anxiety (Sanderson et al., 1989), even though in other blinded studies the CO_2 clearly can trigger panic over and above that set off by sham inhalation

(Woods et al., 1986). At the same time, perceived stress leads to immune dysfunctions in human (Kiecolt-Glaser and Glaser, 1987) and animal (Shavit et al., 1984) subjects. The administration of opiate drugs to opiate-tolerant animals in a strange cage versus a familiar cage doubles the mortality from the same high dose (64% versus 32%; whereas 96% of drug-naive animals die) (Siegel et al., 1982). The mediating mechanisms of these types of effects may involve hormonal, central nervous, and autonomic nervous system elements. Without accounting for the entire clinical situation in MCS, many similar factors might still contribute to the day-to-day course of the syndrome.

Possible Mediators in MCS

In modern health psychology and psychosomatic medicine, it is becoming apparent that the mind/brain and body have a two-way communication system that utilizes biological messengers to implement the effects of "stress" on the body and to change mood and behavior. These include common hormones and other mediators such as cortisol (Sapolsky et al., 1990), interleukins (Sapolsky et al., 1987; Spadaro and Dunn, 1990), vasopressin (Weingartner et al., 1981), substance P, vasoactive intestinal peptide, prostaglandins, kinins, histamine, and opiate-like peptides (enkephalins, endorphins) (Sloviter and Nilaver, 1987; Wiedermann, 1987). For instance, prostaglandins are elevated in major depression (Calabrese et al., 1986; Ohishi et al., 1988), interleukin 1 in allergy and infection (Walter et al., 1989), and opiate-like peptides in sleep apnea (Gislason et al., 1989) and depressive disorders (Agren et al., 1982). PGE2 induces wakefulness (Matsumura et al., 1989); IL-1 produces somnolence and reduces exploratory behavior (Walter et al., 1989; Spadaro and Dunn, 1990); opiate-like peptides facilitate antigen-induced histamine release and lower natural killer cell activity (Shavit et al., 1984; Mediratta et al., 1988).

Many of these same agents participate in more purely physiological events (e.g. inflammation, infection, atopy). However, multiple pathways that originate with biological as well as psychosocial triggers can set off release of these mediators. Once they act at their end-organs, the clinical effects appear similar, regardless of the nature of the original trigger. It is reasonable to hypothesize that many of these mediators contribute to symptom production in MCS from multiple initiating pathways. Clinical symptom pictures in MCS patients (e.g. MFS/MCS daytime somnolence - see Bell et al., 1990 versus somnolence-inducing properties of prostaglandin D2 and interleukin 1) and data on mediators in syndromes overlapping MCS can guide the selection of possible endogenous agents for study.

Classical Conditioning Hypothesis

The hypothesis of classical conditioning of symptom flares to olfactory stimuli represents a potential intersection between psychological and biological mechanisms. For example, antigen exposure in a sensitized animal will mobilize histamine release on a physiological basis. If this event is paired in time with a previously neutral event (e.g. a physiologically-inactive odor), subsequent release of histamine in response to the classically conditioned odor without antigen will occur (Russell et al., 1984; MacQueen et al., 1989). Numerous drug responses including those to opiates and insulin can also be classically conditioned, although the direction of the conditioned response is often opposite from that to the direct biological stimulus (Eikelboom and Stewart, 1982).

In all classical conditioning, it is important to note that the phenomenon requires an initially biologically active stimulus to pair with the previously inactive stimulus. As stated earlier, the existence of classically conditioned responses does not rule out the co-existence of more directly biologically-initiated responses in the same individual. Thus, while it is important to explore classical conditioning of chemical responses in MCS patients and normals (cf. Kirk-Smith et al., 1983; Lorig and Roberts, 1990), the first priority of MCS research must be the biological effects of chemicals. Once the parameters of the biological effects are clarified, then possible psychosocial modifiers and their associated biological mediating mechanisms should be examined experimentally.

METHODOLOGICAL ISSUES

Methodological issues in the study of MCS are critically important. While the scope of this paper is limited to multiple chemical sensitivities, the body of controlled literature on multiple food sensitivities offers valuable clues to the proper design of studies on chemicals and chemically-sensitive patients (Pearson et al., 1983; Egger et al., 1985; Kahn et al., 1988, 1989; Kaplan et al., 1989). Table 2 summarizes the concerns regarding design, subject prescreening, sample size, placebo controls, blinding, duration and timing of avoidance and/or challenges, number of independent measures, and sensitivity of dependent measures. Importantly, the intense debate on the testing and treatment technique known as provocation-neutralization (King, 1988) should be held separately from the more fundamental questions concerning the nature and mechanisms of adverse chemical reactions. For the present, research on MCS should emphasize delivery of real-life-like concentrations via customary routes of exposure (i.e. nose inhalation, dietary ingestion) rather than shifting focus to any particular testing technique or unusual route of administration.

TABLE 2

Methodological Issues		
Issue	Less Sensitive	More Sensitive
design	no control condition	between or within subjects comparison
subject prescreening	all subjects with diagnosis	subset with diagnosis
# subjects	smaller	larger
placebo control	too complex - active	adequate - inactive
blinding	no blind or single blind	double blind
duration of study	single challenge	weeks of challenges
# independent measures	one chemical or food	multiple items
sensitivity of dependent measures	dichotomous global + or - for reaction	continuous variables

From the MFS literature, we have learned that it is necessary to prescreen subjects for current - not merely historical - reactivity to the proposed test agents. Under experimental conditions, researchers have generally not detected effects when studying whole populations - but have found effects when studying subsets of prescreened populations. The latter point raises statistical power considerations in the experimental design (King, 1985). If the effects are limited to a subset of a population, the less certainty we have in identifying these cases on entry into the study necessitates the use of larger sample sizes to avoid Type II error (missing an effect that is actually present by failing to choose enough responders as subjects).

It is also clear that blinding and selection of placebo conditions are difficult issues in a highly sensitive sample of subjects. It would be unfortunately easy to generate a placebo that is physiologically active, though otherwise indistinguishable from the intended active agent. Lorig's research on normal human subjects suggests a possible strategy, which involves a two-step procedure: (i) determination of olfactory thresholds for each test substance in the study subjects, (ii) experiments using sub-threshold concentrations of the test agent (Lorig et al., 1990a,b). This approach avoids sensory awareness and expectations, as well as the complication of possible biological reactivity to added masking odors intended to be inactive while hiding the actual active test agent.

With the background noise of chronic symptomatology and chronic everyday chemical exposures, the experimental approach in MFS also suggests ways to maximize emergence of test reactions under research conditions. That is, investigators need (a) to use repeated, multiple challenges over extended periods of time rather than a single challenge on one day (Kahn et al., 1988, 1989; Kaplan et al., 1989); and (b) to employ multiple rather than single chemical agents for tests (Kaplan et al., 1989). Patients who are in fact chemically sensitive to various substances may nonetheless hold incorrect beliefs about their reactivity to any specific chemical on a single day for a wide range of reasons, including the reality that no chemical exposure in real life is experienced separate from thousands of others in ambient air. Thus, Molhave's approach of starting investigation of tight building subjects and normals with mixtures of volatile organic compounds similar to those found in indoor air may be the most appropriate first step in MCS research as well (Molhave et al., 1986; Molhave 1990; Otto et al., 1990). MCS studies performed in environmentally controlled units with measurably low levels of ambient chemical noise may encounter fewer of these issues of background exposures in their design, especially with subjects resident for weeks at a time.

Finally, it is extremely important to select appropriate and sensitive dependent measures. It is wasteful to perform a controlled study, only to ask subjects whether or not they believe they are having adverse reactions (Pearson et al., 1983). Such dichotomous measures are notoriously insensitive, lower statistical power, and are likely to create Type II error of missing an actual effect of the chemical challenges. In addition to objective laboratory tests such as EEG, evoked potentials (Lorig, 1989), polysomnographic sleep studies (Kahn et al., 1988), blood and cerebrospinal fluid changes in putative mediators and their receptors, and neuropsychological cognition tests (Swanson and Kinsbourne, 1980; Molhave et al., 1986; Lorig et al., 1990b), researchers have a host of standardized self-report and observer-rated scales from medicine and psychology to grade changes in physical and emotional symptoms, behaviors, and mood along a continuum. All such measures offer the possibility of more sensitive continuous rather than dichotomous dependent variables for study.

CONCLUSIONS

In view of the obvious role of the olfactory system in communicating information about the chemical environment to the brain and thereby to the rest of the body, the olfactory-limbic model outlined above is a plausible alternative to the previous emphasis on putative immunological mechanisms for MCS. As indicated, neurons of the limbic brain have excitability properties that could provide the basis for amplification and spreading of adverse reactions to low dose chemical exposures. Furthermore, numerous psychosocial modifiers could participate in the overall clinical course of the illness via their transduced effects on the brain and the bodily functions it regulates. Overall, as is common in the clinical practice of biopsychosocial patient care, the first priority should be to study the direct biological effects of low dose chemical exposures on health in the subset of individuals vulnerable to multiple chemical sensitivity. The next priority is to examine interactions between the biological and psychosocial dimensions of MCS. Eventually, well-done investigations will permit the objective evaluation of the relative proportions of the variance in MCS symptomatology to which biological, psychological, and social factors contribute.

Despite the controversies surrounding multiple chemical sensitivities, it is possible to develop experimentally testable hypotheses concerning the nature and mechanisms of MCS. This work requires synthesis of approaches and information from multiple fields, including occupational medicine and toxicology, internal medicine, health psychology, basic and applied neurosciences, and pharmacology. In this area, it is ultimately insufficient to frame questions in terms of exclusively biological versus psychosocial causes or to show a fit or lack of fit between known clinical entities and MCS. The problems raised by chemical sensitivities demand scientific thought beyond the boundaries of existing fields. The solutions are likely to expand our knowledge base in novel ways toward understanding interactions between human beings and their environment.

REFERENCES

Agren, H., L. Terenius, A. Wahlstrom 1982. Depressive phenomenology and levels of cerebrospinal fluid endorphins. Annals N.Y. Acad. Sci. :388-398.

American Psychiatric Association 1987. Diagnostic and Statistical Manual of Mental Disorders. Third Edition, revised. Washington, D.C.: American Psychiatric Association.

Armstrong, D.L., J.L. Polan-Curtain, T. Osaka, M.R. Murphy, S.Z. Kerenyi, S.A. Miller, S.L. Hartgraves 1990. Central nervous system protection against effects of long-term low-dose nerve agents. Report for USAF School of Aerospace Medicine, Human Systems Division (AFSC), Brooks Air Force Base, Texas (USAFSAM-TR-90-31), U.S. Naval Medical Research and Development Command.

Ashford, N.A., C.S. Miller 1991. Chemical Exposures. Low Levels and High Stakes. N.Y.: Van Nostrand Reinhold.

Bartko, J.J. and S.Kasper 1989. Seasonal changes in mood and behavior. A cluster analytic approach. Psychiatry Res. 28:227-239.

Bell, I.R. 1982. Clinical Ecology: A New Medical Approach to Environmental Illness. Bolinas, CA: Common Knowledge Press.

Bell, I.R. 1987 Environmental illness and health: the controversy and challenge of clinical ecology for mind-body health. Advances 4(3):45-55.

Bell, I.R., Markley, E. J., B. Zumalt, D.S. King 1990. Psychological and somatic symptom and illness patterns in individuals with self-reported adverse food reactions. Biol. Psychiatry 27:168A.

Bell, I.R., M.L. Jasnoski, J. Kagan, D.S. King 1991. Depression and allergies: survey of a nonclinical population. Psychother. Psychosom. 51, in press.

Bentley, S.J., D.J. Pearson, K.J.B. Rix 1983. Food hypersensitivity in irritable bowel syndrome. Lancet 2:295-297.

Black, D.W., A. Rathe, and R.B. Goldstein 1990. Environmental illness. A controlled study of 26 subjects with '20th century disease'. J. Amer. Med. Assoc. 264:3166-3170.

Bokina, A.I., N.D. Eksler, A.D. Semenenko, and R.V. Merkur'yeva 1976. Investigation of the mechanism of action of atmospheric pollutants on the central nervous system an comparative evaluation of methods of study. Environ. Health Perspectives 13:37-42.

Bolla-Wilson, K., R.J. Wilson, M.L. Bleecher 1988. Conditioning of physical symptoms after neurotoxic exposure. J. Occupational Med. 30:684-686.

Cain, D.P. 1974. The role of the olfactory bulb in limbic mechanisms. Psychological Bull.

81:654-671.

Calabrese, J.R., R.G. Skwerer, B. Barna, A.D. Gulledge, R. Valenzuela, A. Butkus, S.Subichin, N.E. Krupp 1986. Depression, immunocompetence, and prostaglandins of the E series. Psychiatry Research 17:41-47.

Cohen, S. and S.L. Syme, eds. 1985. Social Support and Health. New York: Academic Press, Inc.

Dager, S.R., J.P. Holland, D.S. Cowley, D.L. Dunner 1987. Panic disorder precipitated by exposure to organic solvents in the work place. Am. J. Psychiatry 144:1056-1058.

Davidoff L.L., T.J. Callender, L.A. Morrow, G. Ziem 1991. Letter to the Editor. J. Psychosom. Res. 35(4):1-3.

Dilsaver, S.C. 1986. Cholinergic theory of depression. Brain Res. Rev. 11:285-316.

Doty, R.L., D.A. Deems, R.E. Frye, R. Pelberg, A. Shapiro 1988. Olfactory sensitivity, nasal resistance, and autonomic function in patients with multiple chemical sensitivities. Arch. Otolaryngol. - Head & Neck Surgery 114:1422-1427.

Egger, J. C.M. Carter, J. Wilson, M.W. Turner, J.F. Soothill 1983. Is migraine food allergy? A double-blind controlled trial of oligoantigenic diet treatment. Lancet 2:865-869.

Egger, J., C.M. Cater, J.F. Soothill, J. Wilson 1989. Oligoantigenic diet treatment of children with epilepsy and migraine. J. Pediatr. 114:51-58.

Eikelboom, R. and J. Stewart 1982. Conditioning of drug-induced physiological responses. Psychological Rev. 89:507-528.

Engel, G.L. 1977. The need for a new medical model: a challenge for biomedicine. Science 196:129-136.

Gislason, T., M. Almquist, G. Boman, C. Lindholm, L. Terenius 1989. Increased CSF opioid activity in sleep apnea syndrome. Regression after successful treatment. Chest 96:250-254.

Hudson, J.I. and H.G. Pope 1990. Affective spectrum disorder: does antidepressant response identify a family of disorders with a common pathophysiology? Am. J. Psychiatry 147:552-564.

Isaacson, L.G., S.A. Spohler, D.H. Taylor 1990. Trichloroethylene affects learning and decreases myelin in the rat hippocampus. Neurotoxicology Teratology 12:375-381.

James, L., A. Singer, Y. Zurynski, E. Gordon, C. Kraiuhin, A. Harris, A. Howson, R. Meares 1987. Evoked response potentials and regional cerebral blood flow in somatization disorder. Psychother. Psychosom. 47:190-196.

Jamner, L.D., G.E. Schwartz, H. Leigh 1988. The relationship between repressive and defensive coping styles and monocyte, eosinophile, and serum glucose levels: support for the opioid peptide hypothesis of repression. Psychosom. Med. 50:567-575.

Jesberger, J.A. and J.S. Richardson 1985. Animal models of depression: parallels and correlates to severe depression in humans. Biol. Psychiatry 20:764-784.

Jesberger, J.A. and J.S. Richardson 1988. Brain output dysregulation induced by olfactory bulbectomy: an approximation in the rat of major depressive disorder in humans. Intern. J. Neuroscience 38:241-265.

Jones, V.A., P. McLaughlan, M. Shorthouse, E. Workman, J.O. Hunter 1985. Food intolerance: a major factor in the pathogenesis of irritable bowel syndrome. Lancet 2:1115-1117.

Jung, M.W., J. Larson, G. Lynch 1990. Long-term potentiation of monosynaptic EPSPs in rat piriform cortex in vitro. Synapse 6:279-283.

Kahn, A., G. Francois, M. Sottiaux, E. Rebuffat, M. Nduwinana, M.J. Mozin, J. Levitt 1988. Sleep characteristics in milk-intolerant infants. Sleep 11:291-297.

Kahn, A., M.J. Mozin, E. Rebuffat, M. Sottiaux, M.F. Muller 1989. Milk intolerance in children with persistent sleeplessness: a prospective double-blind crossover evaluation. Pediatrics 84:595-603.

Kanter, E.D., L.B. Haberly 1990. NMDA-dependent induction of long-term potentiation in afferent and association fiber systems of piriform cortex in vitro. Brain Research 525:175-179.

Kaplan, B.J., J. McNicol, R.A. Conte, H.K. Moghadam 1989. Dietary replacement in preschool-aged hyperactive boys. Pediatrics 83:7-17.

Katon, W, M.D. Sullivan 1990. Depression and chronic medical illness. J. Clin. Psychiatry 51[6, suppl]:3-11.

Kiecolt-Glaser, J.K. and R. Glaser 1987. Psychosocial moderators of immune function. Annals Behavioral Medicine 9:16-20.

King D.S. 1985. Statistical power of the controlled research on wheat gluten and schizophrenia. Biol. Psychiatry 20:785-787.

King D.S. 1987. The reliability and validity of provocative food testing: a critical review. Med. Hypotheses 25:7-16.

Kirk-Smith, M., C. Van Toller, G. Dodd 1983. Unconscious odor conditioning in human subjects. Biological Psychology 17:221-231.

Levy, S.M. 1985. Behavior and Cancer. San Francisco: Jossey-Bass.

Liskow, B., E. Othmer., E.C. Penick, C. DeSouza, W. Gabrielli 1986. Is Briquet's syndrome a heterogeneous disorder? Am. J. Psychiatry 143:626-629.

Llorens, J., J.M. Tusell, C. Sunol, E. Rodriguez-Farre 1990. On the effects of lindane on the plus-maze model of anxiety. Neurobiology Teratology 12:643-647.

Lorig, T.S. 1989. Human EEG and odor response. Progress in Neurobiology 33:387-398.

Lorig, T.S. and G.E. Schwartz 1988. Brain and odor: I. Alteration of human EEG by odor administration. Psychobiology 16:281-284.

Lorig, T.S., G.E. Schwartz, K.B. Herman 1988. Brain and odor: II. EEG activity during nose and mouth breathing. Psychobiology 16:285-287.

Lorig, T.S., K.B. Herman, G.E. Schwartz 1990a. EEG activity during administration of low-concentration odors. Bulletin Psychonomic Soc. 28:405-408.

Lorig, T.S., E.Huffman, A. DeMartino, J. DeMarco 1990b. The effects of low concentration odors on EEG activity and behavior. J. Psychophysiology 5, in press.

Lorig, T.S., M. Roberts 1990. Odor and cognitive alteration of the contingent negative variation. Chemical Senses 15:537-545.

MacQueen, G., J. Marshall, M. Perdue, S. Siegel, J. Bienenstock 1989. Pavlovian conditioning of rat mucosal mast cells to secrete rat mast cell protease II. Science 243:83-85.

Marshall, P. 1989. Attention deficit disorder and allergy: a neurochemical model of the relation between the illnesses. Psychol. Bulletin 106:434-446.

Matsumura, H., K. Honda, W.S. Choi, et al. 1989. Evidence that brain prostaglandin E2 is involved in physiological sleep-wake regulation in rats. Proc. Natl. Acad. Sci. 86:5666-5669.

McCrae, R.R. and P.T. Costa 1986. Personality, coping, and coping effectiveness in an adult sample. J. Personality 54:385-405.

Mediratta, P.K., N. Das, V.S. Gupta, P. Sen 1988: Modulation of humoral immune responses by endogenous opioids. J. Allergy Clin. Immunol. 81:27-32.

Mesulam, M.M. 1985. Principles of Behavioral Neurology. Philadelphia: F.A. Davis Co.

Molhave, L. 1990. Volatile organic compounds. Indoor air quality and health. Indoor Air 5:15-33.

Molhave, L., B. Bach, and O.F. Pedersen 1986. Human reactions to low concentrations of volatile organic compounds. Environment International 12:167-175.

Nasr, S., E.G. Altman, H.Y. Meltzer 1981. Concordance of atopic and affective disorders. J. Affective Disord. 3:291-296.

Ohishi, K., R. Ueno, S. Nishino, T. Sakai, O. Hayaishi 1988. Increased level of salivary prostaglandins in patients with major depression. Biol. Psychiatry 23:326-334.

Otto, D., L. Molhave, G. Rose, H.K. Hudnell, D. House 1990. Neurobehavioral and sensory irritant effects of controlled exposure to a complex mixture of volatile organic compounds. Neurotoxicology Teratology 12:649-652.

Pearson, D.J., K.J.B. Rix, S.J. Bentley 1983. Food allergy: how much in the mind? A clinical and psychiatric study of suspected food hypersensitivity. Lancet 1:1259-1261.

Post, R.M., D.R. Rubinow, J.C. Ballenger 1984. Conditioning, sensitization, and kindling: implications for the course of affective illness. Pp. 432–466 in Neurobiology of Mood Disorders, R.M. Post and J.C. Ballenger, eds. Baltimore: Williams and Wilkins.

Purtell, J.J., E. Robins, M.E. Cohen 1951. Observations on clinical aspects of hysteria. A quantitative study of 50 hysteria patients and 156 control subjects. J. Am. Med. Assoc. 146:902-909.
Racine, R.J. N.W. Milgram, S. Hafner 1983. Long-term potentiation in the rat limbic forebrain. Brain Research 260:217-234.

Reiman, E.M., M.E. Raichle, E. Robins, F.K. Butler, P. Herscovitch, P. Fox, J. Perlmutter 1986. The application of positron emission tomography to the study of panic disorder. Am. J. Psychiatry 143:469-477.

Rodin, J. 1986. Aging and health: effects of the sense of control. Science 233:1271-1275.

Rosenthal, N.E., Cameron, C.L. 1991. Exaggerated sensitivity to an organophosphate pesticide. Am. J. Psychiatry 148:270.

Roy-Bryne, P.P. and T.W. Uhde 1988. Exogenous factors in panic disorder: clinical and research implications. J. Clin. Psychiatry 49:56-61.

Russell, M., K.A. Dark, R.W. Cummins, et al. 1984. Learned histamine release. Science 216:436-437.

Sanderson, W.C., R.M. Rapee, D.H. Barlow 1989. The influence of an illusion of control on panic attacks induced via inhalation of 5.5% carbon dioxide-enriched air. Arch. Gen. Psychiatry 46:157-162.

Sapolsky, R., C. Rivier, G. Yamamoto, P. Plotsky, W. Vale 1987. Interleukin-1 stimulates the secretion of hypothalamic corticotropin-relasing factor. Science 238:522-524.

Sapolsky, R.M., H. Uno, C.S. Rebert, C:E. Finch 1990. Hippocampal damage associated with prolonged glucocorticoid exposure in primates. J. Neuroscience 10:2897-2902.

Schatzberg, A.F., J.O. Cole 1991. Manual of Clinical Psychopharmacology, 2nd ed. Washington, D.C.: American Psychiatric Press, Inc.

Schwartz, G.E. 1982. Testing the biopsychosocial model: the ultimate challenge facing behavioral medicine? J. Consult. Clin. Psychol. 50:1040-1053.

Shavit Y., J.W. Lewis, G.W. Terman, R.P. Gale, J.C. Liebeskind 1984. Opioid peptides mediate the suppressive effects of stress on natural killer cell cytotoxicity. Science 223:188-190.

Shipley, M.T. 1985. Transport of molecules form nose to brain: transneuronal anterograde and retrograde labeling in the rat olfactory system by wheat germ agglutinin-horseradish peroxidase applied to the nasal epithelium. Brain Res. Bull. 15:129-142.

Siegel, S., R.E. Hinson, M.D. Krank, J. McCully 1982. Heroin "overdose" death: contribution of drug-associated environmental cues. Science 225:733-734.

Simon, G.E., W.J. Katon, and P.J. Sparks 1990. Allergic to life: psychological factors in environmental illness. Am. J. Psychiatry 147:901-906.

Sloviter, R.S., G. Nilaver 1987. Immunocytochemical localization of GABA-, cholecystokinin- vasoactive intestinal peptide-, and somatostatin-like immunoreactivity in the area dentata and hippocampus of the rat. J. Compar. Neurol. 256:42-60.

Spiegel, D., J.R. Bloom, H.C. Kraemer, E.Gottheil 1989. Effect of psychosocial treatment on survival of patients with metastatic breat cancer. Lancet 2(8673):888-91.

Spadaro, F. A.J. Dunn (1990). Intracerebroventricular administration of interleukin-1 to mice alter investigation of stimuli in a novel environment. Brain, Behavior, and Immunity 4:308-322.

Staudenmayer, H. and J.C. Selner 1990. Neuropsychophysiology during relaxation in generalized, universal 'allergic' reactivity to the environment. A comparison study. J. Psychosomatic Res. 34:259-270.

Stripling, J.S., D.K. Patneau, C.A. Gramlich 1988. Selective long-term potentiation in the pyriform cortex. Brain Research 441:281-291.

Sugerman, A.A., D.L. Southern, J.F. Curran 1982. A study of antibody levels in alcoholic, depressive, and schizophrenic patients. Annals Allergy 48:166-171.

Sullivan, M.D., R.A. Dobie, C.S. Sakai, W.J. Katon 1989. Treatment of depressed tinnitus patients with nortriptyline. Ann. Otol. Rhinol. Laryngol. 98:867-872.

Swanson, J.M., M. Kinsbourne 1980. Food dyes impair performance of hyperactive children on a laboratory learning test. Science 207:1485-1487.

Talbott, J.A., R.E. Hales, S.C. Yudofsky, eds. 1988. The American Psychiatric Press Textbook of Psychiatry. Washington, D.C.: American Psychiatric Press, Inc.

Terr, A.I. 1986. Environmental illness: a clinical review of 50 cases. Arch. Intern. Med.

146:145-149.

Tham, R., B. Larsby, B. Eriksson, M. Niklasson 1990. The effect of toluene on the vestibulo- and opto-oculomotor system in rats, pretreated with GABAergic drugs. Neurotoxicology Teratology 12:307-311.

van Riezen H. and B.E. Leonard 1990. Effects of psychotropic drugs on the behavior and neurochemistry of olfactory bulbectomized rats. Pharmac. Ther. 47:21-34.

Walter, J.S., P. Meyers, J.M. Krueger 1989. Microinjection of interleukin-1 into brain: separation of sleep and fever responses. Physiol. & Behavior 45:169-176.

Weingartner, H., P. Gold, J.C. Ballenger, S.A. Smallberg, R. Summers, D.R. Rubinow, R.M. Post, F.K. Goodwin 1981. Effects of vasopressin on human memory functions. Science 211:601-603.

Weidermann, C.J. 1987. Shared recognition molecules in the brain and lymphoid tissues: the polypeptide mediator network of psychoneuroimmunology. Immunol. Lett. 16(3-4):371-378.

Woods, S.W., D.S. Charney, J. Loke, W.K. Goodman, D.E. Redmond, G.R. Heninger 1986. Carbon dioxide sensitivity in panic anxiety. Arch. Gen. Psychiatry 443:900-909.

Neurobehavioral and Psychosocial Aspects of Multiple Chemical Sensitivity

Nancy Fiedler and Howard Kipen

Over the past decade an increasing number of patients have presented with a symptom complex that has been labeled multiple chemical sensitivities or environmental illness. These patients have been characterized to have persistent or recurrent somatic and psychological symptoms attributed to chemical exposures. The symptoms reported and their duration, however, are not consistent with the accepted toxicological properties of the chemicals. Schottenfeld and Cullen (1986), using DSM-III-R diagnostic criteria, initially described such patients to have one of three psychiatric disorders: 1) typical post-traumatic stress disorder; 2) atypical post-traumatic stress disorder; or 3) somatoform disorder. Commensurate with interest expressed in the lay and professional literature, further attempts have been made to characterize these individuals. Cullen (1987) edited an Occupational Medicine State of the Art Review in which he proposed the following relatively rigorous definition: 1) symptoms were acquired in relation to some initial identifiable environmental exposure(s); 2) the symptoms involve more than one organ system (e.g. respiratory and nervous system); 3) the symptoms recur and abate in response to predictable stimuli; 4) the symptoms are elicited by exposures to chemicals of diverse structural classes and toxicologic modes of action; 5) the exposures that elicit symptoms are very low, i.e. many standard deviations below the "TLV" and at levels not known to cause adverse human responses; 6) no single widely available test of organ system function can explain symptoms. This definition was proposed to differentiate MCS patients from patients with post-traumatic stress disorder and somatoform disorder.

Up to this point, subjects reported in the literature have not fully met this more rigorous definition (Brodsky, 1983; Terr, 1986; Bolla-Wilson, et al., 1988; Black et al., 1990;Rosenberg et al., 1990; Simon et al., 1990). For example, Simon et al. (1990) labeled a group of subjects who were currently in workers compensation litigation as chemically sensitive based solely on self-reported lifestyle modifications. Black et al. (1990) reported on psychological profiles of a group of subjects diagnosed as chemically sensitive by physicians practicing clinical ecology. The triggering event for these patients was not uniformly environmental (e.g. stress) nor was the associated diagnosis always chemical sensitivity. For example, some subjects were diagnosed as having chronic candidiasis. Because of the significant regulatory implications of "chemical" sensitivities a need exists to determine if a subset of patients can be found who meet the more rigorous definition proposed by Cullen (1987). If such patients can be identified, it is also important to develop a standardized method for evaluating these patients. Therefore, the purpose of the present paper is to document that this relatively narrow group of patients can be identified and to describe the standardized battery of tests we have chosen to characterize these patients.

CHARACTERISTICS OF MCS PATIENTS

Over the past two years approximately 25 patients have presented at our Environmental and Occupational Health Clinical Center with symptoms suggestive of chemical sensitivity. That is, all patients reported becoming symptomatic to low levels of multiple chemicals/substances encountered in their daily lives. Each patient underwent a comprehensive medical examination by an occupational physician. This consisted of a lifetime and current medical history, review of previous medical records, brief psychiatric history, physical examination, and routine hematology, blood chemistry, thyroid and urine studies. Further testing such as spirometry or chest x-ray were done as needed to document organic conditions which might reasonably explain the symptomatology. While previous organic conditions were often present (e.g. asthma) patients were not excluded if these conditions did not explain the current symptoms.

With regard to psychiatric history, patients who reported the following psychiatric diagnoses on their medical questionnaire or during a medical history were not included in the more narrowly defined group of MCS patients:
1. Evidence of current psychosis, organic brain syndrome, mania, or major depression.
2. History of treatment, within 10 years of onset of symptoms attributed to toxic exposure, for psychosis, organic brain syndrome, hypomania, major depression, somatoform disorder, dissociative disorder, phobia, panic disorder, post-traumatic stress disorder, obsessive compulsive disorder, or personality disorder.

Also, individuals were excluded from the rigorously defined subset of chemically sensitive patients if the following were present: 1) in litigation at the time of the evaluation; 2) in treatment with a clinical ecologist. While such patients may have chemical sensitivity, for various reasons these factors may confound the interpretion of the evaluation measures.

For example, patients involved in litigation or being treated by a clinical ecologist may have altered symptom patterns based on these factors. That is, patients in litigation may have a vested interest in appearing more symptomatic while the treatments or information provided by a clinical ecologist could potentially change a patient's symptom profile.

Based on the above criteria, eleven of the twenty-five patients qualified as presenting symptoms consistent with chemical sensitivity while not having other current or previous conditions or psychosocial factors that could account for their symptoms.

Among those patients who did not qualify, eight did not strictly meet the Cullen (1987) criteria. That is, on careful consideration either they did not have symptoms that wax and wane with exposure or they could not identify a specific point in time or exposure event after which their symptoms began.

Among the remaining six patients, two were involved in litigation, and one had a previous psychiatric history of major depression with electroconvulsive treatment. Three others were not available for further study due to logistic reasons, e.g. moved out of state. Thus, eleven patients met the full criteria proposed by Cullen (1987) and did not have the additional factors outlined previously which could confound an understanding of chemical sensitivity.

Chemically sensitive patients were three men and eight women ranging in age from 28 to 57 with a mean age of 42 for men and 43 for women. Educational level ranged from 12 to 16 years with a mean of 15 years for men and 14 years for women. Two male patients had previously worked in the chemical industry while the third was currently working as a draftsman. Two of the eight women were currently working as secretaries and two were administrative assistants, while the remainder worked as a teacher, bus driver, health educator, and sales clerk. All patients were employed at the time their symptoms began. All of the male patients worked with significant chemical or solvent exposures and first experienced symptoms in response to a change in ventilation or in a process at work, although none experienced a particular episode associated with accidental release or acute intoxication. The identified precipitating exposures and subsequent symptoms for five of the eight women occurred at work. For the remaining female patients, precipitating exposures began outside of work. The nature of the identified precipitating exposures included laying new carpet (N=4), indoor air quality complaints in an office environment (N=2), an adverse reaction to a prescribed drug (N=1), and a home pesticide application (N=1).

PSYCHOSOCIAL AND PSYCHIATRIC EVALUATION

Neurobehavioral Assessment Measures:

The primary neurobehavioral symptoms reported by MCS patients include impaired memory, concentration, and visuomotor speed and coordination. Standardized tests of these functions were selected to assess MCS patients and to compare their performance to a normative group of the same age and sex.

Tests of Memory:

A. California Verbal Learning Test (Delis, et al., 1987) - short and delayed recall of common words with an interference trial:
This test assesses the strategies an individual employs to encode information as well as immediate recall. Therefore, it is sensitive not only to deficits in the amount remembered but also to the way in which material is remembered. It also provides an indicator of short- and long-term memory as well as memory following interference from other tasks. Thus, this test more closely approximates the kind of memory required in day-to-day functioning and the kind of memory deficits often cited among chemically exposed subjects.

B. Digit Span (Wechsler, 1981) - short-term recall of an increasing string of digits:
This test is a part of the World Health Organization's (WHO) core battery of tests recommended for assessment of neurobehavioral effects of toxic chemicals. It has age specific normative data and has been used widely in the psychological and exposure assessment literature. It provides an evaluation of immediate memory for digits which may

require different skills from the California Verbal Learning Test which requires memory for common words.

C.   Wechsler Memory Scale, Visual Reproduction I & II (Wechsler, 1987) short- and long-term memory for designs:

This task requires drawing visual designs from memory and, therefore, assesses the ability to reproduce rather than simply recognize designs.

Tests of Concentration:

A.   Digit Span test to assess concentration:

In addition to memory, this test also provides an assessment of the ability to concentrate.

B.   Stroop Color and Word Test (Trenerry et al. 1989) - determines the ability to concentrate with interference from outside stimuli:
This task is more complex assessing the ability to concentrate when interfering stimuli are presented. As such, it provides a more realistic assessment of concentration. It has been successful in distinguishing patients with brain dysfunction and, therefore, should provide a useful addition to our assessment of dysfunction due to exposure.

Visuomotor Coordination:

A.   Grooved Pegboard Test (Trites, 1981) - determines fine motor coordination in a timed situation:
This task is used frequently as an indicator of fine motor coordination. Savage et al. (1988) as well as other investigators have found significant deficits on this task among exposed subjects.
B.   Digit Symbol (Wechsler, 1981) - requires sustained attention and coordination between eye and hand to record symbols associated with specific digits;
This task is one of the World Health Organization's recommended tasks for the detection of brain dysfunction due to toxic exposure. It is particularly sensitive to brain dysfunction at a minimal level (Lezak, 1983).
        Performance on the above tests of neurobehavioral function can be confounded by differences in age, sex, race, and intelligence (Valciukas, 1985). That is, performance deficits may be due to these variables rather than due to exposures or trauma from exposures. Normative information for the tests cited above has taken into account differences in performance due to age and sex. However, the normative information available generally does not take into account differences in performance due to general intellectual ability. Therefore, a test of verbal ability is also administered as an estimate of overall intellectual ability. It is generally assumed that basic verbal abilities such as vocabulary and well-rehearsed general information will not be affected by acute or transient exposures to toxic substances (Hartman, 1988). Significant discrepancies in performance between a test of reading or vocabulary and those of concentration, memory, and visuomotor skills may help

validate deficits reported by patients clinically. The reading portion of the Wide Range Achievement Test - Revised (Jastak and Wilkinson, 1984) was chosen to assess general word knowledge and as an estimate of overall intellectual ability.

Psychiatric Assessment:

Since many of the symptoms reported by MCS patients may also reflect psychological distress, several standard measures were selected to assess psychiatric symptomatology. The Structured Clinical Interview for the Diagnostic and Statistical Manual of Mental Disorders - Third Edition - Revised (SCID-III-R - DSM-III-R) (Spitzer et al., 1990) is administered to assess current and previous psychiatric symptomatology and diagnoses.

Questions are based on the criteria set forth in the DSM-III-R for the diagnosis of psychotic, mood, substance use, anxiety, somatoform, eating, and adjustment disorders. Evaluation of psychopathology prior to the development of MCS is also important since previous psychiatric status is often cited as a precursor to the development of MCS (e.g. Simon et al., 1990). To serve this purpose, the SCID-III-R interview was modified so that patients are specifically asked if their symptoms had existed prior to the development of chemical sensitivities. Where the SCID-III-R covers only current episodes, it was modified by asking patients if they had any occurrence of the symptoms prior to the development of their current chemically related symptomatology.

Since previous history of somatization disorder is particularly important to consider among these patients, the Diagnostic Interview Schedule-III-A (DIS-III-A) section for somatization disorder is also separately administered to assess lifetime prevalence of somatization disorder (Robins & Helzer, 1985).

The interview format for the DIS-III-A follows a strict set of procedures to ascertain whether physical symptoms reported could be explained medically or by the use of drugs or alcohol. Therefore, this interview format was chosen to be administered by an occupational health nurse with psychiatric experience.

A physician then reviewed all medical explanations given by the patients to ascertain their medical plausibility. Based on this procedure, the number of medically explained and unexplained somatic symptoms, occurring before and after the development of MCS, can be determined.

The Minnesota Multiphasic Personality Inventory (MMPI) (Hathaway & McKinley, 1967) was given to assess current psychiatric symptoms. This self-report inventory has been in use for the past 47 years with a variety of clinical and non-clinical populations. Results from this inventory include validity scales to assess whether the patient responded with a particular response set such as under- or over-reporting of symptoms. If the validity scales are within acceptable limits, then one may proceed to interpret the clinical scales with some evidence that they truly represent the patient.

Psychosocial Functioning:

MCS has also been reported to have a major impact on patient's functioning in daily life (Cullen, 1987). Therefore, several standardized measures of social functioning were also selected. The Psychosocial Adjustment to Illness Survey Self-Report (PAIS-SR) (Derogatis

& Lopez, 1983) was selected to assess the impact that MCS has on the following dimensions of a patient's life: health care orientation, vocational environment, domestic environment, sexual relationships, extended family relationships, social environment, and psychological distress. The scores of MCS patients may be compared to medically ill patients (i.e. diabetics) to contrast the level of social and life functioning. Other chronically ill groups are also available for comparison such as patients with severe coronary artery disease.

The Social Adjustment Scale-Self Report (SAS-SR) (Weissman & Bothwell, 1976) was selected to assess social functioning compared to other psychiatric patients (e.g. depressed or alcoholic patients). This scale gives an overall adjustment score which is based on items covering social adjustment as it pertains to work in or out of home, social and leisure activities, family and marital relationships and economic issues. It has been shown to differentiate depressed patients from normals.

RESULTS AND CONCLUSION

Neurobehavioral Performance:

Chemically sensitive patients did not reveal deficits in concentration or visuomotor skills although some memory tests suggested impairment.

Psychosocial Function:

Psychosocial functioning of MCS patients was impaired relative to diabetics on the PAIS-SR. That is, they had significant disruptions in their social, home and work lives as well as significant disillusionment with the medical care system. On the SAS-SR, they also revealed disruptions in their home and work functioning. These disruptions were comparable to depressed and alcoholic patients.

CONCLUSION

A subset of MCS patients who meet more rigid criteria can be identified and characterized using standardized neurobehavioral, psychiatric, and psychosocial measures of functioning. To make studies comparable, it is important to develop standard selection criteria for patient selection and standardized batteries for patient characterization.

REFERENCES

Black D.W., Rathe A., Goldstein R. Environmental Illness - A Controlled Study of Twenty-six Subjects with 20th Century Diseases. JAMA 1990;264: 3166-3170;#24.

Bolla-Wilson K., Wilson R.J., Bleecher M.L. Conditioning of Physical Symptoms After Neurotoxic Exposure. J. Occup Med. 1988;30:684-686.

Brodsky C.M. Psychological Factors Contributing to Somatoform Diseases Attributed to the Workplace. J Occup Med. 1983;25:459-464.

Cullen M.R. The Worker With Multiple Chemical Sensitivities: An Overview. In: Cullen M, ed. Workers With Multiple Chemical Sensitivities. Philadelphia: Hanley & Belfus Inc; 1987:655-662.

Delis D.C., Framer J.H., Kaplan E., Ober B.A. California Verbal Learning Test Manual. San Antonio: The Psychological Corporation; 1987.

Derogatis L.R., Lopez M.C. The Psychosocial Adjustment to Illness Scale (PAIS and PAIS-SR) Administration, Scoring and Procedures Manual-I., 1983.

Hathaway S.R., McKinley J.C. Minnesota Multiphasic Personality Inventory Manual - Revised. New York: The Psychological Corporation, 1967.

Hartman D.E. Neuropsychological Toxicology - Identification and Assessment of Human Neurotoxic Syndromes. New York: Pergamon Press; 1988.

Jastak S., Wilkinson G.S. Wide Range Achievement Test Administration Manual. Wilmington: Jastak Associates Inc; 1984.

Lezak M.D. Neuropsychological Assessment. New York: Oxford University Press; 1983.

Robins L.N., Helzer J.E. Diagnostic Interview Schedule (DIS) Version III-A. St. Louis: Washington Univ. School of Medicine; 1985.

Rosenberg S.J., Freedman M.R., Schmaling K.B., Rose C. Personality Styles of Patients Asserting Environmental Illness. J Occup Med. 1990; 32:678-681.

Savage A.P., Keefe T.J., Mounce L.M., Heaton R.K., Lewis J.A., Burcar P.J. Chronic Neurological Sequelae of Acute Organophosphate Pesticide Poisoning. Arch of Env Health 1988;43:38-45.

Schottenfeld R.S., Cullen M.R. Recognition of Occupational-Induced Post Traumatic Stress Disorders. J. Occup Med. 1986;28:365-369.

Simon G.E., Katon W.J., Sparks P.J. Allergic to Life: Psychological Factors in Environmental Illness. Am J Psychiatry. 1990;147:901-906.

Spitzer R.L., Williams J.B.W., Gibbon M., First M.B. User's Guide for the Structured Clinical Interview for DSM-III-R. Washington, DC: American Psychiatric Press, Inc.; 1990.

Terr A.I. Environmental Illness: A Clinical Review of 50 Cases. Arch Int Med. 1986;146:145-149.

Trenerry M.R., Crosson B., DeBoe J., Leber W.R. Stroop Neuropsychological Screening Test Manual. Odessa FL: Psychological Assessment Resources, Inc; 1989.

Trites, R.L. Neuropsychological Test Manual. Montreal, Que: Technolab, 1981.

Valciukas, J.A., Control of Confounding Demographic Factors in the Analysis of Neurotoxicological Data. In Environmental Health Document 3. Neurobehavioral Methods in Occupational and Environmental Health. Copenhagen: World Health Organization, 1985.

Wechsler D. Wechsler Adult Intelligence Scale - Revised. New York: The Psychological Corporation; 1981.

Wechsler D. Wechsler Memory Scale - Revised Manual. San Antonio: The Psychological Corporation; 1987.

Weissman M.M., Bothwell S. Assessment of Social Adjustment by Patient Self-Report. Arch Gen Psychia 1976;33:1111-1115.

Acknowledgments

This project was supported by a grant from the Hazardous Substance Management Research Center, A National Science Foundation Industry/University Cooperative Center and A New Jersey Commission on Science and Technology Advanced Technology Center. We wish to acknowledge the contributions of Kathie Kelly- McNeil and Carol Natarelli, who have both been instrumental in providing the evaluation and care of our patients. We also appreciate the consultation of Bernard Goldstein, M.D. and Michael Gochfeld, M.D., Ph.D. in formulating this project. Finally, we appreciate the support services of Joyce Kosmoski and Patricia Hutty in preparing this and other documents for this project.

Diagnostic Markers
Of Multiple Chemical Sensitivity

G. Heuser, A. Wojdani, and S. Heuser

One hundred thirty five patients (75% females) were evaluated for complaints of often disabling sensitivity to small concentrations of multiple chemicals after chemical exposure in the recent or distant past.

A comprehensive evaluation of all subjectively involved systems showed a high yield of abnormal objective findings on random (unrelated to time of exposure) testing. When properly timed, certain immune function tests (TA1 cells and chemical antibody levels) became abnormal, or more abnormal, after unintentional self-reported acute exposure and were thus shown to be potential markers of multiple chemical sensitivity (MCS).

We suggest that appropriate tests of the central nervous system, peripheral nervous system, nose and sinuses, pulmonary function, T-cell subsets, chemical antibodies and autoimmunity be performed. If four of these seven systems show abnormality, the diagnosis of MCS is supported. If certain functions become abnormal or more abnormal after unintentional significant exposure, the diagnosis is confirmed.

INTRODUCTION

The senior author has directed a headache and chronic pain clinic for more than ten years. As clinical histories were taken in increasing detail over the years, it became apparent that in many patients, headaches were triggered by chemicals in the environment. What appeared to be very small concentrations of perfumes, fumes, smoke and mist, would trigger headaches in some patients but not in friends and family members who were present in the same environment. Eventually, the senior author realized that small amounts of chemicals cannot only cause headaches but also a multitude of other complaints in some patients. This realization became a starting point for his increasing interest in chemical sensitivity.

One of the co-authors (A.V.) is also chemically sensitive. This is why, as an immunologist, he became interested in developing appropriate tests for chemical sensitivity. He made chemical antibody testing commercially available to our patients.

The other co-author (S.H.) specializes in computer assisted medical information retrieval. She works one day a week in a one story office which was built more than ten years ago. Furniture, equipment, carpets, and paint are also about ten years old. Never-the-less, she recently developed headaches and nose bleeds. These regularly developed the morning after she had worked in the office. An investigation showed that the manually set air intake valve for that office building had fallen shut. As a result, all air in the office was now being recirculated. No fresh air could enter anymore. When this situation was corrected, both headaches and nose bleeds disappeared. No other person in the office developed headaches and nose bleeds. The conclusion was reached that S.H. is chemically sensitive.

The three of us combined forces in order to develop a team approach to multiple chemical sensitivity (MCS). Consultants in varying specialties are complementing the efforts of our core group.

Initially, we became interested in complaints of chemical sensitivity in our headache patients. Now, we attract an increasing number of environmentally ill patients ("EIs"). Many of them claim disability and request help so that they can get State or Social Security. Others find themselves in litigation (personal injury, workers' compensation). Therefore, an objective evaluation of all patients became mandatory.

Patients with MCS present to their physicians with a startling spectrum of complaints. These complaints are often presented in a manner highly suggestive of a psychiatric disorder. None of them easily lend themselves to measurement. General malaise and weakness, fatigue, headaches, irritability, depression, memory problems, itching, numbness, "creepy and crawly" sensations, burning sensations in the nose, sore throat, hoarseness, shortness of breath, cough, abdominal distress, are all "soft" complaints and therefore not easily documented.

Patients may arrive in the office equipped with oxygen tank, air filter, and mask. Typically, patients describe either one single or repeated exposures to one or more chemicals in the past and the development of chemical sensitivity thereafter. It is striking how similar the history is, regardless of education or walk of life or background. This consistency from one patient to another certainly suggests that we are dealing with a disease entity.

On physical examination there is striking paucity of abnormal findings. Small erythematous lesions over the skin areas exposed to chemicals come and go. On chest examination, wheezing is occasionally found. Soft signs may or may not be present on neurological examination.

Routine laboratory tests (CBC, blood chemistry) and EKGs are also usually normal. We quickly learned that chemically sensitive patients almost regularly refuse to deliberately expose themselves to chemicals. We thus abandoned the idea of exposing our patients to chemicals in a controlled environmental chamber. Instead, we waited for unintentional exposure to occur and suggested testing within a given interval thereafter. The results were then compared with base line values obtained either before or a longer time after acute exposure.

Our results led us to conclude that patients who claim MCS could and should be used as their own controls. Obviously, whenever research funds are available, this patient population should be compared to non-exposed and non-symptomatic matched controls.

A patient with multiple complaints is difficult to evaluate. Most physicians have learned that such a patient usually turns out to suffer from "psychosomatic illness". This is why even a well trained physician is easily persuaded to assign a psychiatric diagnosis to patients with multiple chemical sensitivity. By contrast, it became our goal to look for objective markers of multiple chemical sensitivity. We believe that this paper shows initial steps in this direction.

MATERIALS AND METHODS

Control Population. One hundred sixty healthy volunteers were examined. Their ages ranged from 22 - 55 years. They had no known disease and denied drug use and smoking. 60% were caucasian, 20% african-americans, 20% were hispanics. 62% were females, 38% were males. Their bloods were examined and normal ranges were established for all immune function tests reported in this paper.

Experimental subjects. One hundred thirty five patients (75% females) were selected from our private patient population. All complained of sensitivity to small amounts of chemicals in their work, home, or other (commuting, shopping, hobbies) environments. This sensitivity had diminished the quality of life in most patients many of whom claimed disability. In the extreme they now lived in self-imposed isolation from society. Some had gone to extraordinary efforts to live in an environmentally "clean" home. Some lived in tents on the beach, some in remote locations in California or out of state. Some had held well paying jobs, had made significant contributions to society but had to give it all up because of their environmental illness.

All patients were personally examined by the senior author. Their work or home environments could not be examined by the authors who had to rely on the patients' reports. Material safety data sheets (MSDS) were examined whenever available from the work site.

Table I lists the settings in which patients claimed to have been exposed. Please note that most exposures were due to groups of chemicals rather than a single chemical.

TABLE I

Group of 135 Patients Exposed to Chemicals			
Pesticides	39	Electron Microscopy	1
Sick homes/Buildings	55	Printing	2
Solvents/Fuels	15	Plating	2
Chlorine Gas	1	Art/Wood	3
Copper Compounds	2	Others	15

All laboratory tests were initially obtained at random i.e. regardless of time of chemical exposure or severity of symptoms. Whenever possible, tests were thereupon specifically timed so as to be obtained a given number of days after exposure and resultant symptoms. Patients were thus instructed to wait for follow-up tests until they were involuntarily exposed and developed symptoms as a result.

Whenever symptoms suggested impairment of neurological or psychological function, we suggested an electroencephalogram (EEG) and evoked response studies. MRI of the brain was usually ordered only if these electrophysiological studies were borderline or abnormal. More recently, we started ordering brain mapping and SPECT studies in a select group of patients.

Whenever symptoms suggested a peripheral neuropathy, we ordered nerve conduction and/or neurometric (perception threshold) studies.

If patients specifically complained of nasal or throat symptoms, consultation was obtained from a board certified ENT-specialist. Studies then frequently included sinus x-rays.

Shortness of breath and related complaints led to pulmonary function testing (PFT) and if indicated, chest x-rays.

Patients underwent immunological studies whenever possible.

We routinely obtained CBC and blood chemistry. As indicated, additional studies (viral, bacterial, endocrine and others) were done.

Neurological testing.   Most EEGs were performed in the office of G.H. and included spontaneous sleep, hyperventilation, and photic stimulation. A 16 channel Beckman instrument was used. EEGs had at times been performed elsewhere.

Most MRI brain scans were performed at Medical Imaging Center of Southern California and read by a neuroradiologist.

Perception thresholds were tested with a Neurometer and evaluated by a neurologist with special experience in the use of this instrument. The results are indicators of peripheral nerve (A to C fibers) function.

Evoked response and conduction studies were done and evaluated by board certified neurologists.

SPECT scanning was done at Harbor General Hospital (UCLA affiliated).

Nasal and sinus evaluation. All symptomatic patients were referred to a board certified specialist.

Pulmonary evaluation. Pulmonary function tests (PFT) were done with a Brentwood Spirometer in our office. Some patients had undergone testing elsewhere.

Immune testing. Tests were performed at Immunosciences Laboratories under the direction of A. Wojdani, Ph.D.. In view of the importance of these tests in the context of this evaluation, the methodology is described in detail.

Preparation of formaldehyde-human serum albumin (F-HSA) and Formaldehyde-bovine serum albumin (F-BSA) conjugates:

F-HSA and F-BSA was prepared by the method of Patterson et. al. (1985). Briefly, one mg. of HSA or BSA (Biocell, Carson, Ca.) in PBS (Phosphate Buffered Saline) PH 7.4, each separately, were exposed to 270 of formaldehyde (Fisher Scientific, Fairlawn, N.J.). The mixture was incubated for 30 minutes at 37 C and was then extensively dialyzed against PBS. The F-HSA or F-BSA was sterilized with a 0.2 m filter (Milliport Corp., Bedford, Mass.).

Electrophoretic and immunoelectrophoretic comparison of HSA, BSA with F-HSA and F-BSA was performed to determine conjugation occurrence. Conjugation was evidenced by altered mobility of F-HSA, F-BSA when it was compared with HSA or BSA respectively. Moreover, the number of free amino acid groups present in the F-HSA or F-BSA was determined by the method of Snyder and Sobocinski (1975) and was used to assess the amount of substitution. The number of amino groups bound to formaldehyde was 26 for HSA and 31 for BSA. In this calculation, the formation of intermolecular cross-linking was considered.

Preparation of toluene diisocyanate-human serum albumin (TDI-HSA) and Bovine serum albumin (TDI-BSA) conjugates:

This preparation was similar to the methods of Dewar and Baur (1982). According to this method, 1g HSA or 1g BSA was dissolved in 100 ml of a buffer solution containing potassium chloride (0.05 mol/l), sodium borate (0.05 mol/l), PH 9.4 and cooled to 4 C. Dioxane (10 ml) containing 0.15 ml of toluene diisocyanate was then added dropwise while stirring over a period of 3 hours, followed by addition of 2 ml of ethanolamine, centrifugation, dialysis filtration and lyophilization. Similar to F-HSA and F-BSA, conjugation was confirmed by electrophoresis and determination of free amino groups present in the conjugate. The number of amino groups bound to TDI was 37 for HSA and 43 for BSA. In addition, spectrographic analysis of the conjugate was undertaken according to Zeiss et. al. (1980). There was a marked increase in absorption from 230 to 260 nm which indicated that TDI had become covalently linked to the protein carrier. This increase in absorption did agree with NH_2 group determination only 76% for HSA and 81% for BSA

Preparation of trimellitic anhydride-human serum albumin (TMA-HSA) and Trimellitic anhydride bovine serum albumin (TMA-BSA):

To prepare these conjugates 25 mg. of TMA was dissolved in 0.5 ml of dioxane and added dropwise either to 25 mg of HSA or BSA dissolved in 5 ml of cold 7% $NaHCo3$ in water. After stirring for 60 minutes at 4 C the conjugates were dialyzed against four changes of 0.1 M $NaHCO3$ and one change of buffer. Finally the conjugates were filtered and kept at -20 C until used. OD analyses of TMA-HSA, and TMA-BSA were done to determine the number of TMA residues linked to the corresponding carrier protein. The concentration of the carrier protein was converted to molar concentration with the molecular weight of HSA and BSA. From the ratio of the molar concentration of the TMA ligand and the protein carrier, the ratio of TMA residues per molecules of carrier was calculated. TMA-HSA was estimated to contain 5 TMA residues per HSA molecules and for TMA-BSA seven residues per albumin molecule (Pien et al., 1988).

Preparation of phthalic anhydride-human serum albumin (PA-HSA) and Phthalic anhydride bovine serum albumin (PA-BSA) conjugates:

These hapten-conjugates were prepared by adding 75 mg of PA to a cooled solution of 300 mg of HSA or BSA in 100 ml of H_2O. The reaction mixture was stirred overnight, dialyzed against 0.1 M PBS using tubings with a cutoff of 8000 dalton. Using the method of

Zeiss et. al. (1977) the molar ratios were calculated. The molar ratios were found to be 22/28 for PA/HSA and 25/30 for PA/BSA.

Preparation of benzene ring HSA (B-HSA) and
Benzene ring BSA (B-BSA) conjugates:

For these preparations, 40 mg. of P-aminobenzoic acid was dissolved in 2 ml of 1 N HCL and cooled by immersion in an ice bath. A cold solution of 14 mg/ml was added dropwise. After each addition, the mixture was stirred for 30 seconds. In parallel, one gram of HSA or BSA was dissolved in boric acid 0.16 M sodium chloride (0-15 M) buffer PH 9.0 (PH was raised with NaOH). The beakers containing the solutions of albumins were surrounded by an ice bath on magnetic stirrer. The solution of diazonium salt was added dropwise, with rapid stirring, to the cold protein solution. After addition of each drop the PH is readjusted to 9.0 to 9.5 with one normal NaOH. When all the solution had been added, the reaction was allowed to continue with slow stirring for at least an hour with further additions of NaOH solution and maintaining the PH at the range of 9.0 to 9.5. Unreacted small molecules were removed by extensive dialysis or by passage through a column of sephadex G-25 in the cold room, with an isotonic salt solution as the eluting buffer. OD analyses of orange color development of B-HSA and B-BSA were done to determine the number of B residues linked to the corresponding carrier protein. The amount of B substitution for HSA were approximately 41 and for BSA 53 (Migrdichian, 1957).

Antibody Determinations:

Specific antibodies against F-HSA, TDI-HSA, TMA-HSA, PA-HSA and B-HSA were analyzed by a noncompetitive ELISA assay. Wells of microtiter plates (Dynatech, Alexandria, VA) were coated with 100 l of antigen solutions (100 g/ml) in 0.1 M PBS PH 7.2 overnight at 4 C. Plates were washed 4 times with 0.1 M PBS containing 0.05% tween 20 between each step. Free absorption sites were blocked with 2% protease free bovine serum albumin at room temperature for 4 hours and stored at -20 C until used.

Analytical Procedure:

The procedure included the following: (1) washing four times, (2) addition of 100 l of diluted serum (1:2 for IgE and 1:100 for IgM and IgG) in PBS tween-20 with 1% BSA (3) incubation for 4 hours at 20 C, followed by washing 4 times, (4) addition of 100 l of an optimal dilution of alkaline phosphatase labeled affinity purified goat anti-human IgE ( ) (1:200), IgM (1:500) and anti IgG (1:1000), purchased from KPI (Maryland) (5) incubation for 120 minutes at 20 C, (6) washing 6 times, (7) addition of 100 l of P-nitrophenyl phosphate (Sigma Chemical Co.) (8) incubation for 60 minutes at 20 C (9) addition of 50 l of 3 N sodium hydroxide solution, and (10) duplicate reading. The results were calculated based on absorbances of duplicate samples of 405 nm using microtiter reader. All samples were read against an HSA antigen as a control of
binding not specific to F-HSA, TDI-HSA, TMA-HSA, PA-HSA and B-HSA. Results were

expressed as titer. Titer is being the last dilution of serum giving absorbance twice of HSA control.

Specificity and Cross-inhibition Studies:

For determination of antibody specificity a cross-inhibition study was undertaken. Positive sera for each hapten-protein conjugate were run after appropriate incubation and precipitation with tenfold increasing increments of hapten bound HSA or BSA as inhibitors to cover the range of antibody to antigen excess. This range was between 50 g to 1000 g for hapten-BSA and 80 g to 1000 g for hapten-RSA. After incubation at 37 C and removal of precipitate by centrifugation, the samples from before and after cross-inhibition study were then placed on plates with wells coated with the specific conjugate. The subsequent steps were followed as described above for the ELISA study.

IgG and IgM antibody binding to different conjugates was inhibited by hapten-HSA or hapten-BSA from 36-85%. At a given concentration, both hapten-HSA and hapten-BSA inhibited the antibody level in similar manners.

Partial inhibition of IgE antibody binding to different hapten conjugates was observed when serum was pre-incubated with hapten-HSA or hapten-BSA. This incomplete observation of inhibition of IgE antibody was mainly related to nonavailability of serum with high IgE titers against different chemicals in our laboratory.

Determination of Normal Levels of Antibodies (Controls):

Based on the above procedures, 160 blood donor samples of healthy individuals, of both sexes, between the ages of 22-55, were examined for antibody levels against F-HSA, TD-HSA, TMA-HSA, PA-HSA, and B-HSA. The average titer was 1:800 400 for IgG, 1:3200 1600 for IgM and 1:8 4 for IgE. Thus, in our laboratory titers greater than 1:1600 for IgG, 1:6400 for IgM and 1:16 for IgE are considered positive.

In a given patient, rises or falls in antibody titers by more than one dilution were considered significant (see Tables).

Lymphocyte Subset Enumeration:

A single laser flow cytometer (Epics Profile: Coulter Epics, Inc., Hialeah, FL) which discriminates forward and right angle light scatter, as well as two colors, was used with a software package (Quad Stat: Coulter). Mononuclear cell populations were determined by two-color direct immunofluorescence by using a whole-blood staining technique with the appropriate monoclonal antibody and flow cytometry (Fletcher et al., 1989) The following pairs of fluorescein isothiocyanate (FITC), or phycoerythrin (PE)-conjugated monoclonal antibodies (Coulter immunology) were selected: T11-RDI/B4-FITC, T4-RDI/T8-FITC, T3-FITC/NKH-1-RDI and T11-FITC/Ta1-PE for determination of T-cell/B cell, T-helper/T-suppressor, NKHT3+/NKHY3- and for alternate pathway of lymphocyte activation respectively.

To monitor lymphocyte markers, bit maps were set on the lymphocyte population of the forward-angle light scatter versus a 90 light scatter histogram. The percentage of positively stained cells for each marker pair, as well as the percentage of doubly stained cells was determined. Estimates of absolute numbers of lymphocytes positive for the respective surface markers were determined by multiplying peripheral lymphocyte cell count by the percentage of positively stained cells for each marker pair. Also, the percentage of doubly stained cells was determined. Estimates of absolute numbers of lymphocytes positive for the respective surface markers were determined by multiplying peripheral lymphocyte cell count by the percentage of positive cells for each surface marker.

Measurement of anti-myelin basic protein antibodies:

Human myelin basic protein (HMBP) was prepared by the method of Diebler et al. (1972) and checked for purity by polyacrylamine gel electrophoresis. Antiserum to HMBP was induced in rabbits by repeated injection of HMBP in complete Freund's adjuvant. Antibody activity in the rabbit sera and patient's samples was detected by adding different dilutions (1:100 to 1:10,000) of sera to wells of a microtiter plate previously coated with HMBP as follows: HMBP 250 g/ml was dissolved in carbonate buffer, PH 9.6 and 200 l of this solution were added to each well. After incubation, washing and blocking as above, 200 l of either diluted rabbit or human serum were added to the wells. After incubation for 1 hour at 37 C the sera were shaken out of the wells and then were washed 5 times with wash solution. 200 l of peroxidase-conjugated goat anti-rabbit or goat anti human IgG, IgM or IgA (optimal dilution) were added to the appropriate well. After incubation and repeated washing 200 l of ABTS substrate were added to each well. Plates were incubated for one hour at room temperature and read in a microtiter reader at 405 nm wavelength. Using rabbit antisera, a titration curve was plotted and patient's sera were compared to this standard curve. Based on more than 200 controls and patients' sample determinations, titers greater than 1:2000 for IgA, 1:5000 for IgM, and 1:8000 for IgG were considered positive.

RESULTS

Table II lists the neurological tests done in patients who had complaints of headaches, irritability, memory loss, depression, numbness, tingling, crawling sensation etc. Some studies were done in sufficient number to be suggestive of a significant trend. EEGs were abnormal in 45% of tested patients. They showed mild rather than severe abnormalities, with mostly unilateral (at times bilateral) intermittent slowing, dysrhythmia, and occasional single sharp waves and spikes in the temporal and adjacent leads. MRI scans were also abnormal in a high percentage of cases. In some scans there was a definite impression of atrophy (13%) or demyelinating disease (7%). Others (8%) had more ill-defined non-diagnostic lesions. Visual evoked (VER) and brain stem auditory evoked (BAER) responses were also abnormal in a high percentage of cases.

The number of patients who underwent single photon emission computerized tomography (SPECT) studies of brain perfusion and metabolism as well as computerized analysis of their EEG activity (BEAM) was small but initial results suggest further studies in a greater number of patients.

TABLE II

Neurological Tests in Patients with Multiple Chemical Sensitivity

Test	% Abnormal	# of Patients
Spect	75	4
Conduction	62	13
Neurometer	47	7
EEG	45	76
BEAM	43	7
BAER	33	18
MRI	28	54
VER	25	32

Current perception threshold studies by Neurometer were also performed in only a small number of patients. Never-the-less, they hold promise as potential markers of peripheral neuropathy, just as conduction studies do.

TABLE III

Examination of Nose, Sinuses and Pulmonary Function in Patients with Multiple Chemical Sensitivity

TEST	% ABNORMAL	# of PATIENTS
ENT Specialist	100	19
Sinus x-rays	52	25
PFT	62	78
Chest x-rays	16	32

Table III shows that a thorough ENT exam will show abnormalities in a high percentage of patients. The consistent findings were atrophic rhinitis in patients with severe nasal complaints. Sinusitis or at least thickened mucous membranes were found on sinus examination.

The typical abnormality on PFT was a decrease of FEF 25 - 75% to below 70% of predicted value, indicating small airway disease.

The tests done in table II and III were obtained at random i.e. unrelated to time of exposure.

Table IV shows that, again on random testing, increase in TA1 cell count and percentage is the most frequent abnormality in patients with MCS. Helper/suppressor (H/S) ratios can be increased (50%), unchanged, or decreased upon random testing. Suppressor cells were decreased in 27% of 110 patients. Whether their continued decrease leads to auto-immune disease is not yet apparent from our initial data. Mitogenesis was abnormal in 42% of 12 patients. Normal ranges in our control group were as follows: TA1 Cells 0-432/mm3 or 0 - 10%; H/S ratio 1 - 2.2; Helper Cells 336 - 2,376/mm3 or 35 - 55%; Supressor Cells 192 - 1598/mm3 or 20 - 37%; Lymphocytes 960 - 4,320/mm3 or 20 - 40%; B Cells 48 - 648/mm3 or 5 - 15%.

TABLE IV

Immune Cell Population in Patients with Multiple Chemical Sensitivity

TEST	% ABNORMAL	# OF Patients
TA1	66	92
H/S	50	109
Helper	43	109
Lymphs	27	110
B-Cells	25	104

For normal ranges, see text.

Table V shows abnormal levels of chemical antibodies in a high percentage of patients with MCS. By contrast "normal" (patients unaware of symptoms from chemical exposure) individuals remained in the normal range. (see Materials and Methods section).

Table VI illustrates elevated levels of chemical antibodies in a symptomatic patient. There was a significant change in IgG (benzene ring) and IgM (isocyanate) levels which decreased after exposure ceased and the patient became "asymptomatic" (while however still chemically sensitive).

We also examined for IgE chemical antibodies which were not elevated in any of our studies.

Table VII shows a significant decrease to normal levels of chemical antibodies in a patient who traveled out of state and stayed in a "non-contaminated" environment where she slowly became "asymptomatic". TA1 cells were not a good marker of MCS in this case.

TABLE V

Elevation of Antibodies to Chemicals in Patients with Multiple Chemical Sensitivity		
TESTS	% ELEVATED	# of PATIENTS
All/Any	64	111
TMA	41	111
Benzene	28	74
Isocyanate	30	110
Formaldehyde	30	111
Phthalic Anhydride	6	81

For normal ranges see TABLE VI.

TABLE VI

	SYMPTOMATIC 3-30-90 IgG/IgM	ASYMPTOMATIC 6-22-90 IgG/IgM	NORMAL RANGE IgG/IgM
FO	3,200/1,600	1,600/800	1,600/6,400
ISO	3,200/3,200	1,600/800	1,600/6,400
TMA	6,400/1,600	3,200/800	1,600/6,400
PHTH A	1,600/3,200	800/1,600	1,600/6,400
BENZ	6,400/3,200	1,600/3,200	1,600/6,400

K. W., 42, female, was symptomatic from exposures to auto mechanic repair shop in 3/90. No more exposures after early April 1990.

Table VIII illustrates a case in which chemical antibodies were a poor marker of chemical sensitivity. By contrast, cells of the immune system were significantly abnormal and slowly approached normal as the patient stayed away from her home for several months. While the patient was originally bed ridden from exposure at her home, she was ambulating and much improved after four months away from her home. By contrast, her

TABLE VII

	Before IgG/IgM	+ 4 Mos. IgG/IgM	Normal Range IgG/IgM
FO	6,400/12,800	800/800	1,600/6,400
ISO	6,400/3,200	800/800	1,600/6,400
TMA	3,200/3,200	1,600/3,200	1,600/6,400
TA_1 #	207	168	0-432
TA_1 %	7.6	7.5	0-10

M. D., 53, female, with MCS before and after leaving "toxic" environment at her home.

husband denied chemical sensitivity (he actually had some minor symptoms on detailed questioning) and could only be persuaded to be tested after he was away from his home for four months. It should be noted that the couple re-entered their home every few weeks to fetch some of their belongings. This led to intermittent exposure and may account for the slow recovery of the wife. While the patient claimed exposure to malathion, her home had possibly become contaminated by other pesticides.

Table IX illustrates rapid increase in TA1 cells and decrease in immunocompetent natural killer cells (NKHT3) in a student with MCS who entered an anatomy laboratory for sufficient length of time to become severely symptomatic. By contrast, T-cells and helper-suppressor ratios did not change within that same time interval.

Table X illustrates increase in IgM antibodies to TMA, phthalic anhydride and compounds with a benzene ring, and also in IgG antibodies to the benzene ring, approximately two weeks after significant exposure. By contrast changes in TA1 cells were seen within only one day! Note that other parameters were unchanged when studied during this short time interval.

Table XI illustrates that antibodies to TMA can be the only antibodies elevated after exposure. In table XII only two antibodies (isocyanate and chemicals containing the benzene ring) are elevated after significant exposure. Table XIII illustrates increase in benzene related antibodies in the wife who showed more evidence of chemical sensitivity than the husband. Note that other antibodies were not elevated. Results in tables XI, XII and XIII suggest specificity in the respective chemical antibody tests.

Table XIV shows increased antibodies in husband and wife. Both claimed total disability from severe MCS with the wife showing more symptoms. Testing was done at random and years after the acute exposure. Table XV also illustrates the long term effects of exposure in a mobile home which the couple had moved out of three years earlier. They both claimed MCS and were symptomatic when seen in our office. Note elevated antibodies to formaldehyde and TMA together with high TA1 counts in both husband and wife.

Table XVI depicts the presence of auto-antibodies in our patient population.

TABLE VIII

	Wife + 4 Wks.	Wife + 6 Wks.	Wife + 4 mos.	Husband + 4 mos.	Normal Range
H/S	0.7	0.8	1.1	2.1	1 - 2.2
$NKHT_3$ +#	471	396	77	27	14 - 216
$NKHT_3$ +%	15.1	9.2	3.4	1.7	1,5 - 5.0
TA_1 #	624	689	452	192	0 - 432
TA_1 %	22	16	20	12.4	0 - 10
FO IgG/IgM	800 / 800	----	800 / 800	800 /800	1,600/ 6,400
ISO IgG/IgM	800 / 800	----	800 / 800	800 / 800	1,600/ 6,400
TMA IgG/IgM	800 / 1,600	----	800 / 800	800 / 800	1,600/ 6,400
Phth IgG/IgM	800 / 800	----	1,600 / 1,600	1,600 / 6,400	1,600/ 6,400

Married couple, (male, 45 and female, 56) exposed to malathion with symptoms far more severe in the wife.

Not shown in table XVI are additional results with respect to autoimmunity: ANA titers were positive in 17% of 96 patients. The highest percentage of elevated antibodies was seen when anti-myelin antibodies were studied. They were positive in 80% of 50 patients studied. Typically the elevation was in the IgM and/or IgA rather than IgG antibodies.

In contrast to our findings, "normals" have been reported to have a very low incidence of positive auto-antibodies: ANA 3-4%, parietal approximately 2%, smooth muscle approximately 3%, mitochondrial approximately 1%.

DISCUSSION

We embarked on our studies with the hope that patients with a claim of MCS could eventually be objectively evaluated. We felt that an approach should be found which could accommodate a great number of patients and could be supervised by a primary physician.

Our studies were done by specialists and with equipment available in most cities in the US. The immunological studies described in this paper are sophisticated but can be executed on blood which is mailed over-night to an appropriate laboratory.

The growing number of patients claiming MCS will in our opinion make it impossible

TABLE IX

	BASELINE 7 - 21 - 90	+ 2 DAYS 8 - 14 - 90	NORMAL RANGE
T #	2,975	2,005	701 - 3,758
T %	85.8	85.7	73 - 87
H/S	2.3	2.2	1 - 2.2
TA_1 #	523.7	985	0 - 432
TA_1 %	15.1	42.1	0 - 10
$NKHT_3$ + #	10.4	4.7	14 - 216
$NKHT_3$ + %	0.3	0.2	1.5 - 5

Patient L. W., 34, female with MCS, reacting to exposure in an anatomy laboratory on 8-12-90.

to study them all in environmental chambers under controlled conditions. These chambers will however be needed for further research.

Table XVII shows a list of chemicals brought to the senior author by a patient with MCS. She was exposed to some or all of these chemicals at work on an everyday basis. This table illustrates the dilemma for both patient and physician in trying to attempt to disentangle a complex problem such as this. An environmental chamber approach might not be practical in such a case as it would take too long and would be too expensive to study this matter in great detail.

While exacting research requires well selected controls, these are not easily found in our polluted urban environment. For example, most people are more or less aware of and effected by pollution in Los Angeles. Thus, there seems to be a whole spectrum of sensitivity, with our patient population being at one extreme.

Nevertheless we, have accumulated in excess of one hundred patients who did not complain of MCS and had no elevation of chemical antibodies (also see Material and Methods section for discussion of controls) nor of TA_1 cells.

In view of the above we feel that a patient should serve as his/her own control and that therefore all studies should be longitudinal. Our results show that properly timed studies can bring about significant changes in certain parameters after self-reported exposure. We are not certain at this time when TA1 cell counts reach a peak after exposure. We now know however, that these cells are elevated one to two days after exposure. We are also uncertain when chemical antibody levels reach a peak value. All we have shown so far is an elevation about two weeks after exposure.

Chemical exposure and its effects on the immune system has recently become the subject of discussion by leading allergists (Salvaggio, 1990). Changes in immune cell

TABLE X

Chemical Antibodies and Subpopulations Before and After Exposure

TESTS	9-14-90	10-19-90	TESTS	9-14-90	9-28-90
FO-IgG	800	1,600	WBC	7,300	8,800
-IgM	800	1,600	LYMPH %	27	37
ISO-IgG	800	1,600	T%	76	79
-IgM	800	1,600	H%	47	45
TMA-IgG	1,600	3,200	S%	27	29
-IgM	800	3,200	H/S	1.7	1.6
Phth-IgG	800	800	B%	21	17
-IgM	800	3,200	TA_1%	4.4	16.1
Benz-IgG	800	3,200	TA_1 #	86	512
-IgM	800	6,400			

A. G., Female, 52, MCS from work in sick bldg. Re-entry into same bldg. after severl months on 9-27-90.

populations, specifically TA1 cells, after exposure to chemicals were recently described by another group (Thrasher et al., 1989; Thrasher et al., 1990).

Chemical antibody measurements in chemically exposed patients recently became commercially available. Appropriate immunological procedures were originally developed by Dr. Wojdani and used in patients of the senior author (Thrasher et al., 1987). The original procedures were then expanded by Dr. Wojdani to include additional chemicals (this paper) and adopted and verified by another laboratory (Antibody Assay Laboratories) where additional research was done and published (Thrasher et al., 1989; Thrasher et al., 1990).

At this time, it should be noted that benzene is not per se antigenic. However, our data suggests that some chemical compounds containing the benzene ring are antigenic. Further studies are needed to determine which of these compounds cause antibody formation.

Our data suggest that chemical exposure can push some patients in the direction of autoimmune disease. Multiple sclerosis is an example. A number of our patients are suspected of having that disease on the basis of not only their clinical presentation but also abnormal MRI and evoked response studies together with high anti-myelin antibodies. This was previously discussed (Gard and Heuser, 1990).

Studies other than immune tests should also be done in a longitudinal fashion. Cost containment, a lack of research funding and other factors made this impossible in our patient population. However, PFT were at times studied immediately after exposure and became abnormal. It is possible that some neurological parameters (EEG, BEAM, and SPECT) may also show some significant changes.

TABLE XI

	BASELINE IgG / IgM	+ 2 WEEKS IgG / IgM
FO	1,600 / 1,800	800 / 800
ISO	800 / 800	800 / 800
TMA	1,600 / 3,200	3,200 / 12,800
BENZ	6,400 / 3,200	3,200 / 6,400

Patient, L. W., 34, female, with MCS, reacting to exposure in a "Sick Building".

TABLE XII

	8-14-90 IgG / IgM	10-9-90 IgG / IgM	NORMAL RANGE IgG / IgM
FO	800 / 800	1,600 / 1,600	1,600 / 6,400
ISO	800 / 1,600	3,200 / 1,600	1,600 / 6,400
TMA	800 / 800	1,600 / 800	1,600 / 6,400
PHTH A	800 / 800	1,600 / 800	1,600 / 6,400
BENZ	800 / 800	3,200 / 3,200	1,600 / 6,400

Patient L. C., 45, male, with MCS before and after exposure to "noxious" environment on 9-25-90.

While we have come to expect a high percentage of abnormal immune function tests, we were surprised at the high percentage of abnormal neurological tests. This indicates that the "psychiatric" presentation by many of these patients may well have a neurological basis.

The high number of abnormal test results in our patient population is probably explained by the fact that many patients were disabled with MCS and therefore quite sick.

Our EEG and SPECT studies point toward the limbic system as being involved in MCS. This system's possible role is aptly discussed in this conference by Dr. Miller. It, together with the role of the olfactory system deserves further study (also see Dr. Bell's presentation at this conference). Early studies by Russian authors (Bokina et al., 1976) pointed in this same direction.

TABLE XIII

	WIFE IgG / IgM	HUSBAND IgG / IgM	NORMAL RANGE IgG / IgM
FO	800 / 1,600	800 / 800	1,600 / 6,400
ISO	800 / 1,600	800 / 1,600	1,600 / 6,400
TMA	800 / 1,600	800 / 800	1,600 / 6,400
PHTH A	800 / 3,200	800 / 800	1,600 / 6,400
BENZ	1,600 / 12,800	800 / 6,400	1,600 / 6,400

Married couple (male, 43 and female, 35) with MCS after intermittent exposure (ongoing) to gasoline fumes.

TABLE XIV

	WIFE IgG / IgM	HUSBAND IgG / IgM	NORMAL RANGE IgG / IgM
FO	6,400 / 3,200	6,400 / 1,600	1,600 / 6,400
ISO	12,800 / 1,600	6,400 / 1,600	1,600 / 6,400
TMA	6,400 / 1,600	6,400 / 1,600	1,600 / 6,400
PHTH A	1,600 / 3,200	1,600 / 1,600	1,600 / 6,400

Married couple (male, 38 amd female, 34) with MCS after exposure to roofing materials approximately 2 years earlier.

While EEG studies showed mostly mild abnormalities in the temporal and adjacent leads (see results), our youngest patients (sister and brother, ages two and four respectively) developed actual clinical seizures with grossly abnormal EEGs about three weeks after moving into a new home and playing on the brand new carpet. For several months thereafter the mother observed MCS in both her children. Seizures and MCS slowly abated after the family moved out of the new home.

Diagnostic criteria for MCS. The diagnosis of MCS should be suspected if a patient reports

TABLE XV

	WIFE	HUSBAND	NORMAL RANGE
H/S	2.2	2.0	1.0 - 2.2
TA$_1$ #	923	1,579	0 - 432
TA$_1$ %	45	47	0 - 10
FO (IgG / IgM)	6,400 / 3,200	3,200 / 3,200	1,600 / 6,400
ISO (IgG / IgM)	1,600 / 1,600	800 / 1,600	1,600 / 6,400
TMA (IgG / IgM)	8,000 / 12,800	4,000 / 3,200	1,600 / 6,400

Married couple (male, 60 and female, 58) with MCS after exposure to Formaldehyde in 1986. Tests done in 1989.

TABLE XVI

Autoimmunity in Patients with Multiple Chemical Sensitivity (N = 92)

PARIETAL CELL	24%	MITOCHONDRIAL	4%
SMOOTH MUSCLE	17%	BRUSH BORDER	4%
MICROSOMAL	15%	RETICULIN	3%
THYROGLOBULIN	7%		

impaired well-being whenever exposed to more than one chemical in concentrations which do not effect the general population. In the extreme, the concentrations are very low and the patient is very sick and claims disability.

A comprehensive evaluation of seven systems should then be undertaken. It should be understood that not all seven systems (central nervous system, peripheral nervous system, nose and sinuses, PFT, T-cells subsets, chemical antibodies, autoimmune panel) are always affected. However, we suggest that abnormalities in four out of these seven systems strengthen the suspicion of MCS. If parameters become abnormal or become more abnormal following self-reported acute exposure, the diagnosis is basically confirmed. Studies in environmentally controlled chambers will be necessary to further advance the field.

TABLE XVII

Chemicals Used at Work
Polychlorinated Biphenyls (PCBs)
Pesticides
1,1,1 Trichloroethane
[methyl] Isocyanates
Xylene
N Pentane
Cyclopentane
Hexane Isomers
N-hexane
Formaldehyde
Perchloroethylene
Monochlorodifluoromethane/Chloropentafluoroethane
Barryman blend
Tectyl 802A
PSKD-NF/ZP-9 Developer
SKL-HF/S Spotcheck Penetrant
Nitrocellulose lacquer [Sinclair]
Krylon Enamel
Tapmatic No. 1 cutting fluid
Industrial lacquer
Chevron thinner

The suggestion to use four out of seven criteria is taken from the diagnostic criteria for systemic lupus erythematosus (SLE), where four out of eleven criteria have to be present to make the diagnosis (see table XVIII). It should be noted that psychiatric features are seen in a significant number of patients with SLE. Thus, the same should not be unexpected in patients with MCS.

CONCLUSIONS

Patients who present with complaints of MCS deserve a comprehensive objective evaluation. If this is performed, a high percentage will be shown to have abnormal test results. This is true if the central and peripheral nervous systems as well as pulmonary and immune functions are tested. Also, anatomical changes are frequently found in the nasal passages on close inspection. By contrast, CBC and blood chemistry are usually within normal limits. So are findings on general physical examination.

Whenever possible, longitudinal studies should be performed in which the patient is used as his/her own control. Increases in TA1 cells and chemical antibodies can then be seen following self-reported unintentional exposure and are therefore suggested as markers of MCS.

Our results suggest diagnostic criteria for MCS. These are sorely needed as the number of patients who claim disability as a result of MCS is growing. Millions of dollars are potentially at stake as claims increase. Patients who are truly sick deserve attention and help from industry, housing authorities and government agencies as well as physicians. Patients who make unjustified claims should be quickly identified.

Patients and industry and government are all in need of a practical approach to the diagnosis of MCS. We believe that our findings are pointing the way to such an approach.

Acknowledgements: We thank Dr. D. Alessi for ENT evaluations, Dr. R. Holgate (neuro-radiology) for evaluation of MRI brain scans, Dr. R. Lawrence (neurology) for interpreting the neurometer studies and Dr. I. Mena (nuclear medicine) for evaluation of SPECT brain scans.

REFERENCES

Bokina, A.I., N.D. Eksler, A.D. Semenenko, and R.V. Merkur'yeva. 1976. Investigation of the mechanism of action of atmospheric pollutants on the central nervous system and comparative evaluation of methods of study. Environ. Health Perspect. 13:37-42.

Dewar, M.A., and X. Baur. 1982. Studies on antigens useful for detection of IgE antibodies in isocyanate-sensitized workers. J.Clin. Chem. Clin. Biochem. 20:337-340.

Diebler, G.E., R.E. Martenson, and M.W. Kies. 1972. Large scale preparation of myelin basic protein from central nervous tissue of several mamalian species, Prep. Biochem. 2:139-165.

Fletcher, M.A., S. Azen, B. Adelsberg, G. Gjerset, J. Hassett, J. Kaplan, J. Niland, T. Maryhon, J. Parker, D. Stites, and J. Mosley. 1989. Immunophenotyping in a multicenter study. The transfusion safety study experience. Clin. Immunol. Immunopathol. 52:38-47.

Gard, Z. and G. Heuser. 1990. Re: Multiple sclerosis, solvents and pets. Letter to the Editor. Arch. Neurol. 47:138.

Migrdichin, V. 1957. Conjugated and synthetic antigens in organic synthesis. Reinhold, New York.

Patterson, R., K.E. Harris, and L.C. Grammer. 1985. Canine antibodies against formaldehyde-dog serum albumin conjugates: induction, measurement, and specificity. J. Lab. Clin. Med. 106:93-100.

Pien, L.C., C.R. Zeiss, C.L. Leach, N.S. Hatoum, D. Levitz, P.J. Garvin, and R.Patterson. 1988. Antibody response to trimellitic hemoglobin in trimellitic anhydride induced lung injury. J. Allergy Clin. Immunol. 82:1098-1103.

Salvaggio, J.E. 1990. The impact of allergy and immunology on our expanding industrial environment. J. Allergy Clinical Immunol. 85:689-699.

Snyder, S.L. and P.Z. Sobocinski. 1975. An improved 2,4,6-Trinitrobenzosulfonic acid method for the determination of amines. Ann. Biochem. 64:285-288.

Thrasher, J.D., A. Wojdani, G. Heuser, and G. Cheung. 1987. Evidence for formaldehyde antibodies and altered cellular immunity in subjects exposed to formaldehyde in mobile homes--a brief communication. Arch. of Environ. Health. 42:347-350.

Thrasher, J.D., R. Madison, A. Broughton, and Z. Gard. 1989. Building-related illness and antibodies to albumin conjugates of formaldehyde, toluene diisocyanate, and trimellitic anhydride.
Am. J. Industrial Med. 15:187-195.

Thrasher, J.D., A. Broughton, and R. Madison. 1990. Immune activation and autoantibodies in humans with long-term inhalation exposure to formaldehyde. Arch. Environ. Health. 45:217-223.

Zeiss, C.R., R. Patterson, R., J.J. Pruzansky, M.M. Miller, M. Rosenberg, and D. Levitz. 1977. Trimellitic anhydride-induced airway syndromes. Clinical and immunologic studies. J. Allergy Clin. Immunol. 60:96-103.

Zeiss, C.R., T.M. Kanellakes, J.D. Bellone, D. Levitz, J.J. Pruzansky, and R. Patterson. 1980. Immunoglobulin E-mediated asthma and hypersensitivity pneumonitis with precipitating anti-hapten antibodies due to diphenylmethane diisocyanate exposure. J. Allergy Clin. Immunol. 65:346-352.

Possible Mechanisms
For Multiple Chemical Sensitivity:
The Limbic System and Others [1]

Claudia S. Miller and Nicholas A. Ashford

The limited data available at this time suggest that any mechanism or model that would purport to explain the syndrome of multiple chemical sensitivities would need to address the features most closely associated with this illness (Ashford and Miller, 1991):

1. Symptoms involving virtually any system in the body or several systems simultaneously, and frequently the central nervous system (particularly mood, memory and concentration difficulties).
2. Differing symptoms and severity in different individuals, even among those having the same exposure.
3. Induction by a wide range of environmental agents, including pesticides and solvents.
4. Subsequent triggering by lower levels of exposure than those involved in initial induction of the illness.
5. "Spreading" of sensitivity to other, often chemically dissimilar substances; each substance may trigger a different, but reproducible, constellation of symptoms.
6. Concomitant food and drug intolerances, estimated to occur in a sizeable percentage of those with chemical sensitivities.
7. Adaptation (masking), that is, acclimatization or tolerance to environmental incitants, both chemical and food, with continued exposure; loss of this tolerance with removal from the incitant(s); and augmented response with reexposure after an appropriate interval (for example, 4 to 7 days).
8. An apparent threshold effect referred to by some as the patient's *total load*, a theoretical construct that has been invoked to help explain why an individual develops this syndrome at a particular time. Illness is said to occur when the total load of biological, chemical, physical, and psychological stressors exceeds some threshold for the patient. This concept has emerged from clinical

---

[1]Excerpted from Ashford, N. A. and Miller, C. S. 1991. Chemical Exposures: Low Levels and High Stakes. New York: Van Nostrand Reinhold.

observations; no direct experiments have been done to test its validity in humans; however, animal models do exist. The concept aligns with Selye's work on the general adaptation syndrome (Selye, 1946).

Items 2 and 3 above confound epidemiological investigation of this problem.

Items 3 and 4 suggest a two step process - (1) induction or sensitization resulting from an initial "major" exposure and (2) triggering of symptoms by subsequent lower level exposures to many different incitants.

Although knowledge of the mechanism of a disease may be useful for developing better therapies, such knowledge is not a prerequisite for intervention. Preventing the development of multiple chemical sensitivities in those not yet afflicted may be possible by controlling environmental exposures that cause the initial sensitization.

The most frequently cited physiological theories to explain chemical sensitivity involve the nervous system, the immune system, or the interaction between them because these two systems most clearly link the external environment and the internal milieu.[2] The rapid responsiveness of these systems also makes them attractive candidates because symptoms of food or chemical sensitivity have been reported to develop within seconds of exposure. Many chemicals, such as polybrominated biphenyls (PBBs) and trichloroethylene, affect both the nervous system and the immune system. Until 1980, the idea of a possible direct communication between the nervous and immune systems was widely debated. Subsequently, the existence of a neuroimmunoendocrine axis has been increasingly realized. Several discoveries have helped to confirm the presence of two-way communication between the nervous and immune systems (Payan et al., 1986).

Kilburn proposes that the human nervous system, because it is so highly evolved, may be most susceptible to environmental agents (Kilburn, 1989):

"Sensitivity may be its undoing. The intuitive hypothesis is advanced that the nervous system is the most liable of the body's systems to damage from environmental toxins. Appreciation of damage may be masked because subtle dysfunction is concealed by the nervous system's remarkable redundancy and substitution of functions, or it is overlooked in clinical evaluations which are usually only qualitative."

MECHANISMS INVOLVING THE LIMBIC SYSTEM

The hypothalamus (part of the limbic system) has attracted considerable attention because it is the focal point in the brain where the immune, nervous, and endocrine systems interact (Bell, 1982). Bell notes that assuming a direct cause-and-effect relationship would be premature, but that the hypothalamus could mediate food and chemical addictions in patients with multiple chemical sensitivities. The olfactory system has known links to the hypothalamus and other parts of the limbic system, which has led Bell to speculate that "the olfactory system, hypothalamus and limbic system pathways would provide the neural circuitry by which adverse food and chemical reactions could trigger certain neural, psychological and psychiatric abnormalities." Patients with chemical sensitivities have

---

[2]Discussions of possible immune system mechanisms and psychological hypotheses for chemical sensitivity appear in Ashford and Miller, 1991.

reported food cravings, binges, violence, or hypersexual activity following chemical exposures. A model involving the hypothalamus could help to explain such behavioral changes.

Some authors have alleged that psychological conditioning to odors is responsible for patients' reactions to chemicals. Of course, odor conditioning may occur in selected cases. However, physiological mechanisms involving the limbic system may also occur. A direct pathway from the oropharynx to the brain and hypothalamic and limbic region has been demonstrated in rats (Kare, 1968; Maller et al., 1967). Substances placed in the oropharynx migrated to the brain in minutes via a pathway other than the blood stream and in higher concentrations than if administered via the gastrointestinal tract, suggesting a direct route from mouth (or nose) to brain. Similarly, Shipley showed that inhaled substances that contact the nasal epithelium may cross into the brain and be distributed widely via transneuronal transport (Shipley, 1985). Thus, molecules that are inhaled and contact the olfactory apparatus could influence functions in other parts of the brain.

Ryan and associates studied 17 workers who attributed changes in thought processes, particularly memory and concentration difficulties, or changes in mood to their exposure to solvents (Ryan et al., 1988). Those workers with "cacosmia" (a heightened sensitivity to odors resembling that reported by chemically sensitive individuals) performed most poorly on neurobehavioral tests requiring verbal learning or visual memory. The authors felt their findings supported a hypothesis that chronic solvent exposure may affect the "rhinencephalic structures" (the primitive "smell" brain), the evolutionary precursor of the limbic system.

This phylogenetically ancient part of the brain is present in all mammals. It influences the organism's interaction with its environment in many subtle ways essential for preservation of the individual, its offspring, and the species. *Limbus* (Latin for "margin" or "rim") refers to its appearing like a rim around the edge of the cerebral hemispheres. Figure 1 shows its component parts. Note the close anatomical relationship to the olfactory bulb. The olfactory nerves are the brain's most direct contact with the external environment. Each nerve consists of bipolar neurons, which have one end in the upper part of the nose and the other in the brain (olfactory bulb). Strong odors and even milder ones cause electrical activity in the amygdala and hippocampal areas of the limbic system (Monroe, 1986). Subsensory exposure to chemicals can cause protracted, if not permanent, alterations in the electrical activity of the limbic region, beginning first with the most sensitive structures, particularly that portion of the amygdala that analyzes odors (Bokina, 1976). All parts of the limbic system are intimately interconnected. Interestingly, ablation of the olfactory bulb in laboratory animals serves as a model for depression which investigators have used for testing the efficacy of various antidepressants (Jesberger and Richardson, 1988).

The amygdala is involved in feelings and activities related to self-preservation, such as searching for food, feeding, fighting, and self-protection (MacLean, 1986). The cingulate gyrus appears to influence maternal care and nursing, separation cries between mother and offspring, and playful behavior, including wit and humor (MacLean, 1986). The septum involves feeling and expression relating to procreation. Lesions in the septal area may cause hyperresponsiveness to physical stimuli (such as touching, sounds, or temperature changes), hyperemotionality, loss of motivation, excessive sugar and water intake, and fear of unfamiliar situations (Isaacson, 1982), phenomena reported by some chemically sensitive individuals.

The hippocampus appears important for laying down new memories and thus is essential for learning (Gilman, 1982). Hippocampal lesions may cause difficulty in retaining

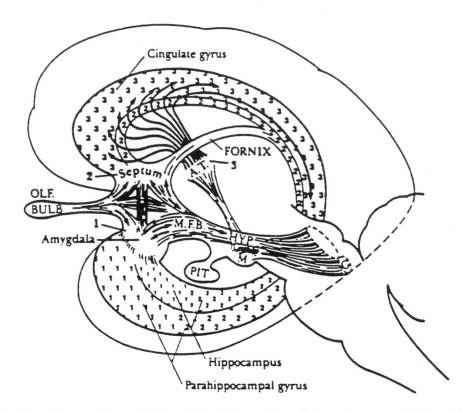

FIGURE 1   Three major subdivisions of the limbic system. The small numerals 1, 2, and 3 overlie, respectively, the amygdaloid, septal, and thalamocingulate divisions. The corresponding large numerals identify connecting nuclei in the amygdala, septum, and anterior thalamus. Abbreviations: AT, anterior thalamic nuclei; G, tegmental nuclei of Gudden; HYP, hypothalamus; M, mammillary bodies; MFB, median forebrain bundle; PIT, pituitary; OLF, olfactory. Source: MacLean, P.D., "A Triune Concept of the Brain and Behavior," in Boag, T., and Campbell, D., The Hincks Memorial Lectures (1973), University of Toronto Press, Toronto, Ontario, p. 15.

recent memories (Isaacson, 1982). The hippocampus, at the intersection of numerous neural pathways and in a critical position to affect the transfer of information from one brain region to another, acts as an information switching center. **Comment:** Learning and memory decrements are a frequent consequence of exposure to toxic substances, and some researchers view the hippocampus as a prime target for such toxins (Office of Technology Assessment, 1990; Walsh, 1988). Damage to the hippocampus itself, or to nerves leading to or from it, may adversely affect the synthesis, storage, release, or inactivation of the excitatory and inhibitory amino acids that serve as neurotransmitters in this region of the brain. Toxins may disrupt the delicate balance of these amino acids, perhaps leading to the release of a flood of excitatory neurotransmitters that damage neighboring cells, a phenomenon that has been called "excitotoxicity" (Office of Technology Assessment, 1990). Remarkably small perturbations of hippocampal function may have large and long-lasting effects upon behavior and cognition (Walsh, 1988).

The most vital component of the limbic system, the hypothalamus, governs: (1) body temperature via vasoconstriction, shivering, vasodilation, sweating, fever, and behaviors such as moving to a cooler or warmer environment or putting on or taking off clothing; (2) reproductive physiology and behavior; (3) feeding, drinking, digestive, and metabolic activities, including water balance, addictive eating leading to obesity, and complete refusal of food and water leading to death; (4) aggressive behavior, including such physical manifestations of emotion such as increased heart rate, elevated blood pressure, dry mouth, and gastrointestinal responses (Gilman, 1982).

The hypothalamus is also the locus at which sympathetic and parasympathetic nervous systems converge. Many symptoms experienced by patients with food and chemical sensitivities relate to the autonomic (sympathetic and parasympathetic) nervous systems; for example, altered smooth muscle tone produces Raynaud's phenomenon, diarrhea, constipation, and other symptoms reported by these individuals.

The hypothalamus also appears to influence anaphylaxis and other aspects of immunity (Stein, 1981). Conversely, antigens may affect electrical activity in the hypothalamus (Besedovsky, 1977).

It is important to recognize that thoughts arising in the cerebral cortex that have strong emotional overtones also can trigger hypothalamic responses and recreate the physical effects associated with intense anger, fear, and other feelings. To implement its effects, the hypothalamus not only has a direct electrical output to the nervous system but also produces its own hormones, many of which stimulate or inhibit the pituitary's production of hormones (Gilman, 1982). Of interest in this regard, is that a disproportionate number of chemically sensitive individuals seem to have been treated for thyroid hormone deficiency at some time in their lives.

Most of the neural input to the hypothalamus comes from the nearby limbic and olfactory areas (Isaacson, 1982). Lesions in the limbic region may be associated with irrational fears, feelings of strangeness or unreality, wishing to be alone, and sadness (MacLean, 1962). A feeling of being out of touch with or out of control of one's feelings and thoughts, not unlike that described by many patients with chemical sensitivity, may be perceived. Some report feeling "spacey" or that "the camera isn't on" unless they make an enormous effort to focus their attention.

Doane describes potential difficulties for patients with limbic dysfunction (Doane, 1986):

> Activity controlled by the limbic system may seem largely irrational and often is not perceived within one's self in ways that are easily understood or communicated in verbal language.

The dynamic involvement of the hypothalamus and limbic system in virtually every aspect of human physiology and behavior makes injury to these structures an intriguing hypothesis to explain the development of chemical sensitivity with its diverse manifestations. Rich neural connections lie between the olfactory system and the limbic and temporal regions of the brain. Surgical or epileptic patients with damage to the limbic or medial temporal portions of the brain may experience persistent alterations in odor perception (for example, an unusual smell that characteristically precedes seizure activity) as well as learning and memory difficulties (Ryan, 1988).

Bell hypothesizes that chemically sensitive patients may have olfactory-limbic-temporal pathways that are more easily "kindled" (Bell, 1990). In other words, a small signal or insult would more readily trigger firing of nerve cells in brain regions where kindling was present.

Kindling might be enhanced by genetic endowment, prior environmental exposures, psychological stress, hormonal variations, or other factors. Unlike surgical ablation, which destroys a brain area, kindling is a kind of stimulatory lesion (Girgis, 1986).

Kindling has been described previously in the context of seizures. The amygdala, for example, which is particularly susceptible to electrical discharge following either electrical (Girgis, 1986) or chemical provocation (Bokina, 1976), is subject to long-lasting alteration when strong or repetitive stimuli are administered. Very potent or repeated stimuli, whether electrical or chemical, may permanently augment the tendency for neurons to fire in the presence of future stimuli, even when challenged with much lower levels than those originally involved.

Girgis reports a decrease in acetylcholinesterase (AChE), an enzyme that breaks down the neurotransmitter acetylcholine in junctions between nerve cells, that parallels the increase in supersensitivity to stimuli (Girgis, 1986). The limbic system is especially rich in AChE, which is strongly bound to the nerve cell membranes and very stable. The AChE may play a protective role by enzymatically maintaining acetylcholine concentrations at nerve junctions within safe bounds and protecting susceptible cells in the limbic system from developing "bizarre sensitivity" (Girgis, 1986). Interestingly, physicians who treat patients with multiple chemical sensitivities have noted some of the most severe and debilitating exposures for these patients have involved organophosphate pesticides, which inhibit AChE.

Bokina found impaired speed of execution and coordination of execution and complex motor processes in humans repeatedly exposed to carbon disulfide for 10- to 15-minute intervals at subsensory levels (Bokina, 1976). Animals primed by high or chronic lower concentrations of various chemicals, such as formaldehyde and ozone, and subsequently reexposed to even lower concentrations of the same chemicals showed an increased tendency toward paroxysmal electrical discharge in the amygdala (Bokina, 1976). Bokina observed that although the chemicals he used to sensitize the animals were different in terms of their structure and physical and chemical properties, their effects upon the limbic system were remarkably similar.

Kindling could help to explain the apparent loss of adaptive capacity and spreading of sensitivities to chemically unrelated substances reported in multiple chemical sensitivity. Formerly well-tolerated low-level exposures to, for example, tobacco smoke or perfume might trigger symptoms in individuals whose limbic areas have been kindled by a prior pesticide or solvent exposure.

One intriguing aspect of the limbic system as a mechanism for multiple chemical sensitivities is the system's responsiveness to both chemical and cortical stimuli. Therefore, conscious thought processes and emotional states influence limbic activity just as chemical or physical stimuli can. The former may be under more or less conscious control of the individual, whereas the latter are almost entirely unconscious and automatic. However, conscious efforts that play into the delicate circuitry of the limbic system may be able to alter or suppress concurrent electrical activity evoked by environmental agents. Some patients with chemical sensitivities report being able to "will" their way out of a mild reaction to a food or chemical and attempt to control their symptoms in this manner. Most say such efforts do not work for their most problematic incitants. In fact, the ability to exercise any conscious effort, even that of simply getting away from the exposure, may be lost during a reaction. Monroe reported the case of a man for whom exposure to the odor of stale beer caused greatly increased electrical activity in the limbic system (amygdala and hippocampal areas). Various memories, some associated with beer, also increased electrical activity in the same region. However, simple arithmetic computations would immediately stop such activity. Therefore, conscious thought processes could alter some electrical activity in the limbic system (Monroe, 1986).

An intriguing example of the competing effects of exposure and psychological state has been reported (Sanderson, 1989). Carbon dioxide levels greater than 5 percent in the air have been shown to induce panic attacks ("fight or flight" responses depend upon limbic activity) in patients suffering from panic disorder. The fact that patients in this study who believed that they had control over the carbon dioxide level to which they were exposed had fewer and less intense panic disorder symptoms suggests that psychological factors (that is, the illusion of control), could mitigate the biological response to an environmental stressor.

Thus, experimental evidence suggests a delicate interplay occurring in the limbic region. Conceivably, chemicals contacting olfactory nerve projections in the nose could either be transported into or relay electrical signals to the limbic region, leading to a vast array of symptoms. Likewise, thought processes and mood states trigger limbic activity or may, in some cases, interrupt preexisting limbic activity. At present, however, there is no evidence to suggest that limbic activity triggered by environmental exposures can be entirely overcome by psychological interventions.

One important ramification of a limbic hypothesis, if true, is that there may be no convenient biological marker for multiple chemical sensitivity.

Detection of chemical stimuli in the nose is not limited to the olfactory nerve, but involves the trigeminal nerve and its afferents which may also play a role in this condition. Trigeminal free nerve endings in the nose and mouth detect noxious chemicals and reflexively initiate protective responses including cessation of breathing, constriction or dilatation of the airways, reduction in heart rate and cardiac output, constriction of most blood vessels (except capillaries in the head), increased epinephrine release, changes in blood pressure and efforts to withdraw (Silver and Maruniak, 1981). This is a powerful reflex which serves an obvious protective role. Most familiar is the trigeminal reflex response to smelling salts. Apparently trigeminal responses can occur with non-irritating stimuli. Thus far, no odor has been found that stimulates the trigeminal nerve alone or the olfactory nerve alone (Silver and Maruniak, 1981) making study of either system in isolation difficult. Potentially both play a role in chemical sensitivity.

BIOCHEMICAL MECHANISMS

Rea and other ecologists who see chemically sensitive patients have noted vitamin and mineral abnormalities in many of them. Others argue that these patients are often sick, debilitated, and malnourished, and therefore, such findings are not surprising.

Individuals who have defective enzyme detoxification systems could be more susceptible to low level exposures. Conceivably, chemically sensitive individuals could have defective detoxification pathways, and be more affected while others in the same environment tolerate the same exposures without symptoms (Rogers, 1990). Rea has noted that many of his chemically sensitive patients have decreased levels of detoxifying enzymes, such as glutathione peroxidase. This possibility is particularly intriguing because such enzyme systems are inducible (that is, can be stimulated) and thus might conform to an adaptation hypothesis. Scadding and associates noted poor sulfoxidation ability in 58 of 74 patients with well-defined reactions to foods versus 67 of 200 normal controls p <0.005) (Scadding et al., 1988). Similarly, Reidenberg reported the case of a laboratory technician who developed a lupuslike disease in response to hydrazine (Reidenberg et al., 1988). She was genetically a slow acetylator, which may might have predisposed her to developing a lupuslike disorder after sufficient exposure to the chemical. A deficiency of one or more particular enzymes could help to explain why some patients are more susceptible to foods and chemicals than others. Further, damage by a toxin might compromise detoxification pathways so that other substances formerly metabolized by this pathway could not be degraded properly and thus might provoke symptoms at low exposure levels, a hypothetical basis for the spreading phenomenon.

Levine has proposed that environmental sensitivities are the result of toxic chemicals reacting with cell constituents to create free radicals (which are formed when a molecule loses an electron) (Levine, 1983). He hypothesizes that if an antioxidant molecule (such as vitamin A, C, E, or selenium) is not present nearby to supply the missing electron, then an electron may be removed from an unsaturated lipid in a cell membrane, leading to membrane damage, release of prostaglandins and other inflammatory mediators, and formation of antibodies to chemically altered tissue macromolecules.

VASCULAR MECHANISMS

Rea hypothesizes that blood vessel constriction, inflammation, or leakage in multiple organ systems could explain the bizarre combinations of symptoms in these patients. In his view, particular complaints could mirror the site and size of affected blood vessels. Spasms in large-caliber arteries, either acutely or chronically, could reduce blood supply to an organ or limb and result in dysfunction, pain, or even necrosis (Rea, 1975). Chemical injury to the fragile walls of smaller vessels, however, would be more likely to cause hemorrhage (resulting in petechiae and bruises) or edema (Rea, 1979). The walls of blood vessels contain smooth muscle. Rea notes that other tissues containing smooth muscle such as the respiratory, gastrointestinal, and genitourinary systems are frequently involved in these patients (Rea, 1977). Impaired blood vessel or altered smooth muscle function might explain the diverse and seemingly unrelated symptoms occurring in patients with multiple chemical sensitivities. In the case of either blood vessel or smooth muscle dysfunction, clearly, neurological and immune alterations or changes in membrane permability or receptors (for example, denervation - supersensitivity) could play primary roles. A vascular hypothesis might also explain why some patients experience increased pain or other symptoms at the site of an earlier injury or surgery, where blood flow may be relatively compromised.

Perhaps the mechanism for multiple chemical sensitivities is not identifiable; that is, after all avenues of biochemical and immunological inquiry have been exhausted, no single explanation for this disorder is forthcoming. The theory of substance-specific adaptation is based upon observations of the responses of patients in a deadapted state who are worked up in an environmental unit. Adaptation is *only* an observation at this time, not a mechanism. However, biological *limits* might regulate how much an organism can adapt, limits that could be highly individual and vary by orders of magnitude. Certainly adaptation occurs at all levels of biological systems, from enzyme systems to cells, tissues, organs, and even behavior (Fregly, 1969). Theoretically, a major insult or the accumulation of lower-level injuries within these systems could lead to a kind of "overload" or "saturation" effect with respect to adaptive capacity that would cause an individual to have environmental responses, which, instead of being flexible and fluid, are now fragile and overly responsive. Many patients report that even after years and in some cases decades following the onset of their problems they have recovered only a portion of their former energies and tolerance for their environment. Their descriptions seem to suggest the loss of an intangible capacity to adapt, parts of which may be temporary and recoverable and other parts of which may not.

REFERENCES

Ashford, N. A. and Miller, C. S. 1991. Chemical Exposures: Low Levels and High Stakes. New York: Van Nostrand Reinhold.

Bell, I. R., Clinical Ecology. 1982. Common Knowledge Press: Bolinas, California.

Bell, I. R. 1990. The biopersonality of allergies and environmental illness. Paper presented at the Eighth Annual International Symposium on Man and His Environment in Health and Disease. Dallas, Texas.

Besedovsky, H., Sorkin, E., Felix, D. and Haas, H. 1977. Hypothalamic changes during the immune response. Eur. J. Immunol. 7:323-325.

Bokina, A. I. Eksler, N. D., Semenenko, A. D., and Merkuryeva, R. V. 1976. Investigation of the mechanism of action of atmospheric pollutants on the central nervous system and comparative evaluation of methods of study. Environ. Health. Perspect. 13:37-42.

Doane, B. K. and Livingston, K. E., editors. 1986. Clinical psychiatry and the physiodynamics of the limbic system. Pp. 285-315 in The Limbic System: Functional Organization and Clinical Disorders. New York: Raven Press.

Fregly, M. J. 1969. Comments on cross-adaptation. Environ. Res. 2:435-441.

Gilman, S., and Winans, S. W. 1982. Manter and Gatz's Essentials of Clinical Neuroanatomy and Neurophysiology. Philadelphia: F. A. Davis Company.

Girgis, M. 1986. Biochemical patterns in limbic system circuitry: biochemical-electrophysiological interactions displayed by chemitrode techniques. Pp. 55-65 in Limbic System: Functional Organization and Clinical Disorders. New York: Raven Press.

Isaacson, R. L. 1982. The Limbic System. New York: Plenum Press.
Jesberger, J. A. and Richardson, J. S. 1988. Brain output dysregulation induced by olfactory bulbectomy: an approximation for the rat of major depressive disorders in humans. Intern. J. Neurosci. 38:241-265.

Kare, M. 1986. Direct pathways to the brain. Science. 163:952-953.

Kilburn, K. H. 1989. Is the human nervous system most sensitive to environmental toxins? Arch. Environ. Health. 44(6):343-344.

Levine, S. A. and Reinhardt, J. H. 1983. Biochemical pathology initiated by free radicals, oxidant chemicals, and therapeutic drugs in the etiology of chemical hypersensitivity disease J Orthomolecular Psychiatry. 12:166-183.

MacLean, P. D. 1967. The brain in relation to empathy and medical education. J Nerv

Ment Dis. 144(5):374-382.

Maclean, P. D., Boag, T. and Campbell, D. 1973. A Triune Concept of the Brain and Behavior: The Hincks Memorial Lectures. Pp. 6-66. Toronto, Ontario. University of Toronto Press.

Maclean, P. D. 1986. Culminating developments in the evolution of the limbic system: the thalamocingulate division. In Doane, B. K. and Livingston, K. E., editors. Pp. 1-28. Organization and Clinical Disorders. New York: Raven Press.

Maller, O., Kare, M. R., Welt, M. and Behrman, H. 1967. Movement of glucose and sodium chloride from the oropharyngeal cavity to the brain. Nature 213(2):713-714.

Monroe, R. R. 1986. Episodic behavioral disorders and limbic ictus. Pp. 251-266 in The Limbic System: Functional Organization and Clinical Disorders. Doane, B. K., and Livingston, K. E., editors. New York: Raven Press.

Office of Technology Assessment, Congress of the United States. 1990. Identifying and Controlling Poisons of the Nervous System. Washington, D. C.: United States Government Printing Office.

Payan, D. G., McGillis, J. P. and Goetzl, E. J. 1986. Neuroimmunology. Adv. Immunology. 39:299-323.

Rea, W. J. 1977. Environmentally triggered small vessel vasculitis. Ann. Allergy. 38:245-51.

Rea, W. J. 1979. The environmental aspects of ear, nose and throat disease. Part I. Oto-Rhino-Laryngology & Allergy Digest 41(7):41-56.

Rea, W. J., Bell, I. R. and Smiley, R. E. 1975. Environmentally triggered large-vessel vasculitis. Pp. 185-198 in Allergy and Medical Treatment. Johnson, J. and Spencer, J. T., editors. Chicago: Symposia Specialists.

Reidenberg, M. 1983. Lupus erythematous-like disease due to hydrazine. Am. J. Med. 75:363-370.

Rogers, S. A. 1990. A practical approach to the person with suspected indoor air quality problems. Indoor Air. Fifth International Conference on Indoor Air Quality and Climate, Toronto, Ontario 5:345-349.

Ryan, C. M., Morrow, L. A. and Hodgson, M. 1988. Cacosmia and neurobehavioral dysfunction associated with occupational exposure to mixtures of organic solvents. Am. J. Psychiatry 145(11):1442-1445.

Sanderson, W. 1989. The influence of an illusion of control on panic attacks induced via inhalation of 5.5 percent carbon dioxide-enriched air. Arch. Gen. Psychiatry 46:157-162.

Scadding, L. 1988. Poor sulphoxidation ability in patients with food sensitivity. Br. Med. J. 297(6641):105-107.

Selye, H. 1946. The general adaptation syndrome and the diseases of adaptation. J. of Allergy. 17:231-247, 289-323, 358-398.

Shipley, M., 1985. Transport of molecules from nose to brain: transneuronal antegrade and retrograde labeling in the rat olfactory system by wheat germ agglutinnin-horseradish peroxidase applied to nasal epithelium. Brain Res. Bulletin. 15:129-142.

Silver, W. L. and Maruniak, J. A. 1981. Trigeminal chemoreception in the nasal and oral cavities. Chemical Senses 6(4):295-305.

Stein, M., Keller, S. and Schleifer, S. 1981. The hypothalamus and the immune response. Pp. 45-65. In Weiner, H.,Hofer, M. A. and Stunkard, A. J., editors. Brain, Behavior and Bodily Disease. New York: Raven Press 45-65.

Walsh, T. J., and Emerich, D. F. 1988. The hippocampus as a common target of neurotoxic agents. Toxicology. 49:137-140.

Developing Clinical Research Protocols for Studying Chemical Sensitivities

William J. Meggs

The decade of the eighties brought progress in identifying and quantifying the low level exposures to volatile organic compounds found in the indoor air of homes, schools, and offices. In addition, some data is now available on both endogenous and added low molecular weight organic chemicals in food and water. Claims of adverse health effects from these exposures are far-ranging, and the challenge is now upon the clinical investigator to scientifically determine if there are adverse health effects associated with these exposures. There are specific difficulties associated with the development of research protocols in this area, and issues that must be addressed include: (a) Determination of patient populations to be studied, (b) the determination of proper control groups, (c) proper techniques for blinding challenges with odorous substances, (d) assessing psychiatric problems, (e) allowing for "adaptative" phenomena, (g) designing studies which will simultaneously establish if stimulus-response relationships exist while investigating possible underlying mechanisms and perhaps most difficult, (f) designing studies that will be definitive in a highly controversial and politicized area of investigation.

There are three approaches to defining patient populations to be studied. In the first circumstance, patients are recruited on the basis of reactivity to chemicals, either by meeting a case definition for multiple chemical sensitivity syndrome or claimed reactivity to multiple chemicals, with the focus on the etiological agents which may or may not reproduce the syndrome. In the second approach patients are recruited that have specific diseases such as asthma, rheumatoid arthritis, or depression, with effects of withdrawing and reexposing patients to specific chemicals determined. In the third design, a given population such as a group of workers in a specific factory are sampled for response. In all three approaches, studies must be designed employing objective parameters to measure disease activity. Studies on patient populations with polysomatic complaints and no objective findings is not likely to yield definitive data. Patients with both objective illness and alleged chemical reactivity are available and are preferred patients for study.

Diseases are potentially induced or exacerbated by environmental chemicals if: (a) the disease incidence has increased in recent years, (b) occupational exposure induces or exacerbates the disease, (c) there are case reports or anecdotal data relating the disease to chemical exposures, (d) the disease activity is linked to pharmaceuticals, as in drug-induced autoimmune hemolytic anemia, and (e) if animal models of the disease show induction by chemical exposure. Using these criteria, a group of diseases that need to be studied is given in Table 1. We do not claim that all of these diseases rigorously meet each of the above criteria. Common features of these diseases are a waxing and waning course, localized tissue inflammation, and ultimately tissue destruction. They are candidates for Stage 2 and Stage 3 illnesses of the Hypothetical Chemical Stress Syndrome, discussed elsewhere in these proceedings.

TABLE 1

Some of the Diseases Whose Relationship to Environmental Chemicals Needs Clarification

Asthma

Autoimmune hemolytic anemia

Crohn's Disease

Depression

Manic depressive illness

Multiple sclerosis

Rheumatoid Arthritis

Schizophrenia

Systemic Lupus Erythematosis

Ulcerative Colitis

Clinical research protocols must be designed to incorporate suitable control groups. Investigators must recognize that the entire population is exposed to low-level volatile organic chemicals (VOC's), and there is data suggesting that normal volunteers exposed in challenge chambers to mixtures of VOC's can become symptomatic. Some effects of chronic low-level VOC exposure may be seen in both normal individuals and patients.

Clinical studies of adverse effects of chemicals must be double-blinded, but unique difficulties arise in the case of odorous inhalants. Odor masking by exposing the subject to a high dose chemical considered innocuous and a low dose chemical to be tested presents unique problems in this setting and will require systematic study. Anesthetizing the olfactory bulb may mask olfactory mediated physiologic reactions. The experience of blinding techniques in the study of food allergy and food intolerance should be heeded, and these investigators have found that blinding techniques should not be accepted a priori, but should be subjected to scientific scrutiny. Exposures of patients during sleep may be the most effective technique, but will need controlled investigation due to altered physiology during sleep.

Clinical studies must allow for psychiatric illnesses in the patient population. It is recognized that there are behavioral aspects to this illness. In the original description of the multiple chemical sensitivity syndrome, it was claimed that psychiatric symptoms including depression, hallucinations, and manic states can be induced in some patients by chemical exposure. A contrary view is that this population consists of psychiatric patients with somatization and chemical phobias. Clinical studies can address this controversy by assessing the psychiatric state of patients in the presence and absence of chemical exposures. Great care must be employed in the design of double-blinded psychiatric evaluations.

In the original clinical description of the multiple chemical sensitivity syndrome, it was claimed that chemically sensitive patients chronically exposed to chemical inhalants do not react acutely to chemical challenges and go through a withdrawal phase lasting approximately five days when removed from the chemical environment. Chronic symptoms are said to resolve after withdrawal, but the patients develop acute symptoms when rechallenged with chemicals. Clinical studies must allow for this adaptation mechanism in order to be accepted. Considerations of possible adaptive phenomena will require residential clinical research units in which indoor air can be controlled. A further consideration during the period of study includes monitoring for reactivity to both endogenous and added organic chemicals in food, and contaminants of water.

Study design should include specific laboratory studies to objectively address defined questions. Though the pathogenesis of this syndrome has not been established, proposed mechanisms should be considered and used as a basis to design a laboratory panel to be followed as patients enter the protocol, undergo withdrawal from the chemical environment in a clinical research unit, and then are reexposed to specific agents. For instance, Are there changes in helper and suppressor lymphocyte numbers, or lymphocyte activation, during this process? Following acute challenge, are substances such as histamine or metabolites of neurotransmitters released into the blood or urine?

In summary, scientific understanding in this area of clinical medicine will only develop with the establishment of clinical research units designed to minimize volatile organic chemicals in the air, staffed with multidisciplinary teams of specialists, and employing well-designed clinical protocols. Patients should be studied in defined populations such as those characterized by quantifiable, objective diseases. Disease activity and laboratory parameters should be followed from entry through withdrawal from organic chemicals, and then upon reexposure. Trials must be double-blinded, with careful assessment for changes in disease activity and in laboratory parameters. Normal volunteers should undergo identical evaluations. Costs of this research and potential sources of funding will be discussed.

Immunological Mechanisms of Disease
and the Multiple Chemical Sensitivity Syndrome

William J. Meggs

INTRODUCTION

The Multiple Chemical Sensitivity Syndrome (MCS) is a highly controversial illness which afflicts an unknown number of people. The very existence of this syndrome is questioned, but it is clear from the experience of physicians from several specialties that a population of patients exists that goes to great lengths to avoid a group of odorous organic chemicals at levels that are known to occur in contemporary homes, schools, and offices. These chemicals include the products of combustion, cleaning products, fragrances, pesticides, outgassing from synthetic materials used for clothing, furnishing, and building products and are commonly found in the indoor air.

This type of sensitivity is colloquially called "allergy" but differs from traditional allergic reactions. At the present time the mechanism of this intolerance of chemicals is as controversial as the syndrome itself, but it is important to consider hypothetical mechanisms at this early stage of scientific scrutiny. Some say there is not scientific evidence that this syndrome exists, but the syndrome exists because the patient population exists, and in extreme cases these people are disabled and no longer able to function productively in society. What is expressed by disclaimers is a belief that the mechanism of the syndrome is a mental disturbance such as somatization, and at this stage the somatization hypothesis is one of several to be considered by investigators designing clinical research protocols.

This article will discuss proposed and hypothetical immunological mechanisms for the Multiple Chemical Sensitivity Syndrome. In the next section, descriptions of this syndrome will be discussed, and the syndrome will be described in terms of a chemical stress syndrome. Known immunological mechanisms for the pathophysiology of established diseases will then be discussed and compared with MCS to see if any are feasible explanations. Finally, the hypothesis that the mechanism of MCS involves neuroimmunology, the interplay between the immune and nervous systems, will be presented.

DESCRIPTIONS OF THE
MULTIPLE CHEMICAL SENSITIVITY SYNDROME

An *opinion* that exposure to environmental organic chemicals could lead to illnesses was expressed in 1962 in a short book, *Human Ecology and Susceptibility to the Chemical Environment*, written by a Chicago allergist (Randolph, 1962). The characterization of this illness is presented in Table 1, and this type of sensitivity was presented as being distinct from classical allergic sensitization. Notable features of this description include an adaptation phenomenon, with: (a) sensitive individuals having chronic symptoms while living in the chemical environment which resolve several days after removal from the chemicals, (b) withdrawal symptoms upon removal, and (c) "shock reactions" when these individuals are reexposed to chemicals. Patients affected by this disorder have a variety of physical and *mental* illnesses which can be managed by avoiding incriminated chemicals. A spreading phenomenon was described, in which affected individuals become progressively sensitive to more and more chemicals found in indoor air, water, and foods. A set of chemicals which were found in clinical practice to cause illness in some patients is given in Table 2. This work was presented as an opinion based upon the clinical practice of having patients avoid chemicals and foods using trial and error, and no scientific data such as double-blinded studies were presented to substantiate this point of view.

TABLE 1

Randolph's Description of Chemical Sensitivity Disease

Acquired, often after chronic insidious or acute exposure to a petrochemical.

Chemical exposures trigger physical (arthritis, asthma, colitis, etc.) or mental (depression, difficulty with concentration, mania, psychosis, etc.) symptoms.

Specific Adaptation Syndrome. Adaptation to specific chemicals with chronic exposure is followed by maladaptation and chronic illness, withdrawal symptoms when removed from the chemical environment, and "shock reactions" on reexposure.

Spreading phenomena. As an individual becomes maladapted to the chemical environment, intolerance to increasing numbers of environmental chemicals develops.

Avoidance. By avoiding the chemical environment, chronic illnesses may resolve.

A case definition of the Multiple chemical Sensitivity Syndrome based on the experience of occupational medicine specialists around the country has been presented (Cullen, 1989). Table 3 gives the seven criteria presented in this case definition. Comparison of Tables 1 and 3 shows considerable overlap, and a discussion of the relationship between the two will be given below.

An operational definition of MCS has been given (Ashford and Miller, 1989, 1991),

TABLE 2

Chemicals Alleged to Cause Reactions in chemically Sensitive Individuals (From Randolph, 1962)

Indoor air contaminants
  utility gas
  coal smoke
  fumes from fresh paint, turpentine, mineral spirits, detergents
  fragrances from toiletries and perfumes
  cleaning products such as bleach, ammonia, disinfectants
  inspect sprays and repellants
  odorous synthetic carpets, pads, adhesives, and building materials

Outdoor air contaminants
  automobile and diesel exhaust
  industrial air pollutants
  paint manufacturing and sulfur-processing fumes
  fumes from roof tar and roads

Chemical additives and contaminants of food and water
  insecticides and fumigant residues
  some chemical preservatives
  sulfur residues
  chemical flavoring and sweetening agents
  plastic containers and lined tins
  chloride in water

Synthetic drugs, cosmetics, and miscellaneous chemicals
  medications, including aspirin, sulfonamides, synthetic vitamins
  allergic extracts or other biologic materials that contain phenol
  cosmetics
  synthetic textiles
  bed linens washed with detergents, dried in gas driers, or impregnated with
  synthetic starch or sizing

that "The patient with multiple chemical sensitivities can be discovered by removal from the suspected offending agents and by rechallenge, after an appropriate interval, under strictly controlled environmental conditions. Causality is inferred by the clearing of symptoms with removal from the offending environment and recurrence of symptoms with specific challenge."

A description of this syndrome is best given in the context of what we call the "Hypothetical Chemical Stress Syndrome", a concept which is not original and draws heavily on the work of others (Selye, 1946; Randolph, 1962). The word "hypothetical" is used because at the present time there is insufficient data to establish a consensus about the role

TABLE 3

Examples of Host Responses that Must Also Be Considered Immunological

Endothelial cell reactions

Activation of C cascades in the absence of antibody

Activation of procoagulant activity

Acute phase reactant responses

Aracidonic acid metabolism

NK cell enhancement

Lysosomal enzyme release

Generation of toxic oxygen radicals

IL-1 effect on the hypothalamus

Beta-endorphin activity

Generation of neuropeptides

of this syndrome in human disease. The word "chemical" is used because our focus is on the stress of exposure to environmental chemicals, though some would argue that emotional and other stresses can "cross react" with chemical exposures.

The stages of the Hypothetical Chemical Stress Syndrome (HCSS) are given in Figure One. Individuals in Stage 0 tolerate chemical exposures without difficulty. Progression from Stage 0 to Stage 1 occurs when an individual is chemically stressed either by an acute high-dose chemical exposure, or by a chronic insidious exposure. Individuals in Stage 1 have symptoms on exposure to chemicals, but no physical findings on physical examination. Hence, Stage 1 is the "-algia" stage, with -algia being the Greek suffix used in medicine to denotes pain; e.g., arthralgia denotes joint pain while myalgia denotes muscle pain.

Stage 2 is reached when an individual develops foci of inflammation from chemical exposures. It is at this stage that both findings on physical examination appear and a medical diagnosis can be given. The specific medical diagnoses which might be amenable to this approach are discussed elsewhere in these proceedings (Meggs, 1991). This stage is referred to as the "-itis" stage, with -itis being the Greek medical suffix denoting inflammation. Arthritis is inflammation of joints, myositis is inflammation of muscle, and so forth. The progression from Stage 1 to Stage 2 again follows increasing chemical exposures, and if tissue damage has not occurred, the inflammation can be reversed by removal of the chemical stimuli. That is, progression between Stages 1 and 2 is a two way process, with progression from Stage 2 to Stage 1 being possible if chemicals are avoided. The inflammation of Stage 2 can be reduced by medications such as corticosteroids and the non-steroidal anti-inflammatory agents, but these medications are not curative. If the chemical stimuli are not removed, there is immediate relapse of inflammation. Further, these medications do not prevent the progression from Stage 2 to Stage 3.

Stage 3 occurs if an individual suffers damage to tissue, so this stage can be

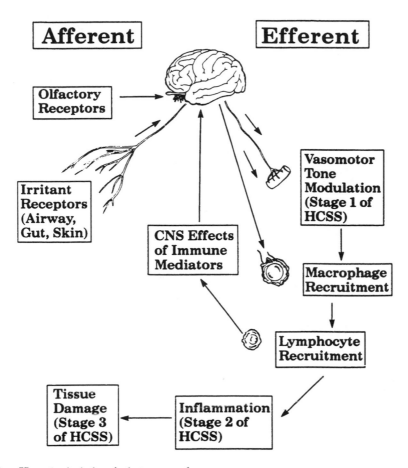

FIGURE 1   Hypothetical chemical stress syndrome.

characterized by "-osis".   In medicine, the Greek suffix -osis denotes a change in living
tissue; e.g., necrosis means death of tissue and fibrosis means the replacement of functional
tissue by scar tissue.   Unfortunately, once tissue is damaged there is little hope in current
medical practice for a reversal, and organ function is lost.

Generally, the progression into Stages 2 and 3 occurs in one organ system, such as the
skin, joints, or lungs.   Later, a second organ system may become involved.   It may still be
important to remove an individual who has reached the irreversible tissue damage of Stage 3
from the chemical environment because it may be that inflammation can occur in other
tissues.

What many clinicians call the Multiple Chemical Sensitivity Syndrome is Stage 1 of the
Hypothetical Chemical Stress Syndrome.   These patients have multiple complaints which
they attribute to chemical exposures, but little if any findings on physical examination.
Hence, it is easy to regard these patients as having a somatization disorder with no real
medical illness.   It is important that studies of adverse reactions to environmental chemicals
include patients with diagnosable medical conditions, that is Stage 2 and 3 patients, and that
the hypothesized progression between these stages be studied.   In this regard, longitudinal
studies will be very important.

The concept of a stress syndrome was originally introduced as the "Generalized

Adaptation Syndrome" to explain observations in animal models of disease (Selye, 1946). In this syndrome, laboratory animals exposed to stresses progress through stages of adaptation similar to the stages described above. The syndrome was initially discovered while exposing animals to chemical stresses, and the word "generalized" was introduced in the name of this syndrome because other stresses, for example the stress of confinement, was observed to lead to similar results.

The "Specific Adaptation Syndrome" has been described in humans to differ from the "Generalized Adaptation Syndrome" in that maladaptation can occur to one specific stress, in particular a single chemical (Randolph, 1962). The extent to which these stress syndromes are equivalent can be debated, but in any event they are very similar. The purpose of formulating the Hypothetical Chemical Stress Syndrome is to develop a specific hypothesis for studying adverse reactions to environmental chemicals such as those found in the indoor air.

IMMUNOPATHOGENESIS OF DISEASE
AND CHEMICAL SENSITIVITIES

It is well known that the interaction of the immune system with environmental chemicals can cause disease. In this section, the known immunological mechanisms of disease will be reviewed, and their possible relationship to the multiple chemical sensitivity syndrome will be discussed. The function of the immune system is to protect the organism from invasion by pathogens such as bacteria, fungi, viruses, and parasites. It is known that the immune system has different arms, or branches, as listed in Table 4, which are recruited to deal with different types of pathogens. We know that these branches are distinct and recent investigations suggest that activation of one branch can actually deactivate another (Moore et al., 1990).

TABLE 4

Branches of the Immune System and Their Function

Branch	Function
IgE-eosinophil system	Destruction of tissue parasites
IgG-neutrophil/macrophage system	Destruction of encapsulated bacteria
lymphocyte system	Destruction of fungi and viruses
IgA-secretory system	Destruction of pathogens at mucosal surfaces

Further, activation of the immune system can lead to inflammation, disease, and tissue damage. IgE mediated reactions arise when foreign proteins or low molecular weight compounds such as drugs bound to foreign proteins interact with IgE bound to the surface of mast cells to release histamine and other immune mediators, leading to the classical allergic reactions to foods, aeroallergens, pharmaceuticals, and insect venoms. Clinical manifestations of these reactions are asthma, urticaria, rhinitis, anaphylactic shock, or gastrointestinal problems. Some of the mediators released in allergic reactions are chemotactic factors, that is, they attract white blood cells into the tissues. Some patients with MCS report flushing and itching after chemical exposures, and a minority of patients report that they developed classical allergic disease after the onset of MCS. These anecdotes do not change the fact that MCS is distinct from classical allergy, as originally stated by Randolph.

Immune complexes are high molecular weight aggregates of antigen and antibody which can deposit in vessel walls, activate complement, and lead to inflammation and tissue damage, with serum sickness being the classical and experimental form of immune complex disease. The clinical picture of serum sickness is distinct, with fever, urticaria, fatigue and malaise, arthritis, and sometime renal failure. Immune complexes have been associated with autoimmune diseases and a host of other clinical conditions (Cochrane, 1988). While they do not play a role in MCS, they may play a role in stage 2 and 3 illnesses of fig. 1 related to chemicals.

Delayed type hypersensitivity, also sometimes referred to as cellular immunity and mediated by T-lymphocytes, can lead to inflammatory changes in tissues after a sensitizing exposure followed by reexposure. The initial exposure leads to antigen-specific T-cells which become activated on reexposure and release cytokines. Experimentally this is most commonly demonstrated in the skin, and contact dermatitis in humans arising from organic chemicals is mediated by cellular immunity. Of the known types of immune reactivity, cellular immunity is the one most likely to play a direct role in adverse reactions to environmental chemicals. The delayed or withdrawal nature of the inflammatory symptoms described in these patients is similar to the delay seen in cellular inflammation. Two of the interleukins (IL-1 and IL-2) are known to have effects on the central nervous system (Obal et al., 1990 and De Sarro et al., 1990). Several authors have claimed abnormalities of cellular immunity in these patients, such as alterations of helper and suppressor ratios and the presence of activated lyumphocytes in peripheral blood, as discussed below.

There are at least theoretical analogies in immunology to the spreading phenomena discussed for patients with multiple chemical sensitivities. The concept of an environmental adjuvant has been introduced to describe substances which induce an immune response to other substances. In immunology, an adjuvant is a substance which potentiates an immune response to other substances. An example of an adjuvant is alum, for an animal injected with alum mixed with a protein will begin producing IgE specific for the protein. Examples of inhaled environmental adjuvants in experimental models include sulfur dioxide, which can induce allergic asthma to co-administered aeroallergens in a guinea pig model (Riedel et al, 1988), and ozone, which has been shown to potentiate platinum sensitivity in monkey (Biagini et al., 1986).

The mechanism of environmental adjuvancy may be that inflamed tissue processes antigen. Substances like sulfur dioxide and ozone produce tissue irritation, which then leads to inflammation with the recruitment of macrophages to the area. Macrophages then present antigens in the area to lymphocytes to produce an immune response to the antigen. Subsequent exposure to either the adjuvant or the antigen could produce further inflammation and processing of other antigens. The process may be initiated by the interaction of chemical irritants with sensory nerves.The repertoire of triggers of inflammation would grow, with increased risk of more substances becoming triggers. Such an immunological spreading phenomena has not been demonstrated to my knowledge but is theoretically possible.

REVIEW OF SELECTED PUBLISHED DATA
ON IMMUNE ABNORMALITIES
IN CHEMICALLY SENSITIVE PATIENTS

Studies have appeared which attempt to establish an immunological basis for patients with multiple chemical sensitivities, as described in Table 5. There are studies depicting increases in T4/T8 ratios in chemically sensitive patients, decreases in this ratio, and no abnormalities in this ratio. It is important to note, however, that the patient populations differ in these studies. One study described three groups of chemically sensitive patients, those with vascular dysfunction (vasculitis), asthma, and rheumatoid arthritis (Rea et al., - +1986). Chemical sensitivity was verified by chemical challenges in these study patients. Rea's patients would be considered stage 2 and 3 patients in the Hypothetical Chemical Stress Syndrome discussed in section I above. Statistically significant decreases in T8 (suppressor) cells and increases in the T4/T8 (helper:suppressor ratio) were found in patients with vasculitis and rheumatoid arthritis but not in asthmatics.

Levin and Byer (1989) have looked at populations of patients who have had documented environmental exposures to chemicals such as trichlorethylene exposure, PCB's, or a mixture of industrial dyes, solvents, and pesticides. These patients were said to be chemically sensitive following these exposures, and may represent Stage 1 patients in the HCSS. This study found decreased T4:T8 ratios in these patients.

Terr examined the medical records of 50 patients referred by the California Worker's Compensation Board of Appeals who claimed to have difficulties with workplace chemicals and found no consistent pattern of immunological abnormality. In contrast to the patients studied by Rea and his collaborators, these patients had a variety of medical diagnoses and no attempt was made to challenge them with chemical exposures to see if they were sensitive to chemicals. Rather, it was concluded that, since another medical diagnosis could be given, they did not have chemical sensitivities and instead had a somatization disorder.

These three studies obtain conflicting results on the T4:T8 ratio in chemically sensitive populations, but the patient populations differ dramatically. Rea's group studied patients with known inflammatory diseases which improved when isolated from the chemical environment, Levin and Byer studied populations environmentally exposed to organic chemicals, and Terr studied patients claiming chemical sensitivity who had had a claim for worker's compensation rejected. These studies illustrate the importance of using well-designed clinical research protocols to assess chemically sensitive patients, using well-defined patient populations. It can be argued that the expression multiple chemically sensitivity is best used as an etiology, not as a medical diagnosis. The hypothesis presented in Section II is that a number of inflammatory conditions can arise from the chronic stress of exposure to environmental chemicals, and it may be that the various conditions developing from chemical exposures may have different immunological profiles reflecting the medical diagnosis rather than the chemical etiology.

One group of investigators has proposed that antibodies to formaldehyde coupled to

TABLE 5

Selected Published Data on Immune Abnormalities in Chemically Exposed Patients

Reference	Patient Group	Number	Abnormality
Rea et al., 1986	Vasculitis	70	T8 decreased[a], T4/T8 increased[a]
	Asthma	27	T4% decreased[b]
	Rheumatoid arthritis	7	T8 decreased[a], T4/T8 increased[c]
Terr, 1986	Worker's comp. appeals	50	No immunological abnormality
Broughton & Thrasher, 1988	Formaldehyde exposure	67	Antibodies to formaldehyde-albumin Increased CD26 cells
Levin & Byers, 1989	TCE exposure	25	T4/T8 decreased[d]
	Industrial Chemical Exposure	10	T4/T8 decreased[e]
			T4/T8 decreased[e]
	PCB exposure	21	T4/T8 decreased[e]
	Electronic workers	78	
Patterson et al., 1989	Chemically sensitive		No antibodies to formaldehyde-albumin
	Dialysis patients		Antibodies to formaldehyde-albumin
McConnachie & Zahalsky, 1991	Chlordane/heptachlor	27	Increased CD26, decreased CD45R/T4 Decreased ConA mitogen response Decreased mixed lymphocyte response

a $p < 0.0001$
b $p < 0.05$
c $p < 0.001$
d $p < 10^{-8}$
e $p < 10^{-6}$

human serum album (f-HSA) may be a good marker for chemical sensitivity (Broughton and Thrasher, 1988; Thrasher et al., 1988; Thrasher et al., 1989). These investigators found low level IgG, IgE, and IgM titers to f-HSA, in the 1:4 to 1:16 range, in chemically sensitive patients. Well-established associations between diseases and antibody titers, such as the association of systemic lupus erythematosis with anti-DNA titers, are characterized by extremely high titers of 1:1,000 or higher. Non-specific binding at the high serum concentrations found in low titer wells make low-titer associations with disease of dubious clinical significance. Another group has found no association between formaldehyde conjugates and chemical sensitivity, but has found titers to be a possible marker of exposure to formaldehyde in renal dialysis patients (Patterson et al., 1989).

An increased frequency of the CD26 lymphocyte marker in chemical sensitivity individuals has been reported (Broughton and Thrasher, 1988). This marker has also been reported with greater frequency in chlordane/heptachlor exposed individuals (McConnachie and Zahalskyu, 1991), though the issue of chemical sensitivity is not addressed in this study. Since CD26 is a lymphocyte activation marker, the notion that chronic activation of cellular immunity by chemical exposures is an attractive one which deserves further study.

NEUROIMMUNOLOGY AND CHEMICAL SENSITIVITIES

Over the past decade there has been an increased appreciation of the numerous interactions between the nervous system and the immune system. It is tempting to speculation that an understanding of chemical sensitivity may be found in the study of neuroimmunology, given the neurological and behaviorally abnormalities in this patient population and the altered reactivity to normally harmless levels of environmental chemicals reminiscent of immune mediated phenomena. In this section we propose a hypothetical model of disease based on the interactions of the immune and nervous systems, as depicted in Fig. 2. Odorous chemicals excite the nervous system in two ways, through olfactory nerves and irritant nerve fibers in the airway. According to the hypothetical tenet of neuroimmunology, given in Table 6, stimulation of irritant fibers in the airway by volatile organic chemicals leads to inflammation. Normal volunteers exposed for four hours to a mixture of volatile organic chemicals were found to have increased numbers of polymorphonuclear cells in nasal lavage washings eighteen hours (Koren et al., 1990), suggesting an association between irritant chemicals and inflammation.

TABLE 6

The Hypothetical Tenets of Neuroimmunology

Tenet 1. Irritancy leads to inflammation and vasomotor abnormalities.

Tenet 2. Inflammation provokes an immune response, that is inflamed tissue processes antigen.

Tenet 3. Immune mediators modulate the central nervous system.

As the airway becomes inflamed, macrophages are recruited to the area and begin to process antigens, including organic chemicals bound to tissue proteins, for presentation to lymphocytes. We propose that this process leads to sensitized lymphocytes specific for the chemicals which were present in the airway. Subsequent exposure to these chemicals can

HYPOTHETICAL CHEMICAL STRESS SYNDROME

STAGE 0
Normal
Tolerates Chemical Exposures

STAGE 1
Irritancy
(-algias)
Symptoms Dominate Signs

STAGE 2
Inflammation
(-itis)
Signs Predominate

STAGE 3
Tissue Damage
(-osis)

FIGURE 2

lead to the production of cytokines which can have systemic effects and may effect the central nervous system. Two interleukins, IL-1 (Obal et al., 1990) and IL-2 (De Sarro et al., 1990) are known to have activity on the central nervous system. Perhaps the difficulty with concentration that is reported by chemically sensitive individuals following exposures is related to the release of interleukins with central nervous system activity.

Blood flow in vessels throughout the body is modulated by the contraction and relaxation of smooth muscles in the blood vessel walls, or vasomotor tone, and the tone of these muscles is controlled by the nervous system. Tenet Number 1 in Table 6 proposes that stimulation of irritant receptors by chemicals can alter blood flow by dilating or constricting vessels. Clinical manifestations would include headache, nasal congestion, and rhinorrhea. It is interesting to speculate if neuronal pathways can be acquired so that stimulation by chemicals in the airway might lead to vasomotor phenomena in other organs. One would then have the possibility of protean manifestations of chemical exposures depending on one's "learned" or acquired responses.

It is instructive to discuss the Hypothetical Chemical Stress Syndrome of fig. 1 in terms of the neuroimmunological mechanisms proposed in fig. 2. Stage 1 patients in the HCSS have an exaggerated vasomotor response to the stimulation of irritant receptors in the airway, with vasomotor tone alterations in organs other than the airway. Stage 2 patients are those who have developed recurrent foci of inflammation resulting from chemical exposures. Stage 3 arises when tissue is destroyed by the inflammatory products including free radicals released by macrophages and neutrophils.

SUMMARY

We have presented a speculative but comprehensive and mechanistic model to explain the alleged adverse reactions to environmental chemicals. The multiple chemical sensitivity syndrome must be discussed in the broader context of stress syndromes. The patient with multiple complaints related to chemicals but few objective signs of disease manifests an early stage of chemical induced disease. We believe that the stimulation of irritant receptors by chemicals, primarily in the airway, leads to modulations of vasomotor tone. In the chemically sensitive patient, this response is exaggerated and can occur various organs through acquired neuronal pathways.

As the disease progresses, foci on inflammation develop, with infiltrations of macrophages, polymorphonuclear cells, and perhaps lymphocytes into the affected tissue. At this point, the patient develops a diagnosable disease and is traditionally treated with corticosteroids, non-steroidal anti-inflammatory agents, bronchodilators, or other pharmaceuticals as appropriate. The inflammation can be reversed by avoiding the causative chemicals if tissue damage has not occurred.

The inflammatory response leads to processing antigen and specific immunological responses to chemicals. Immune mediators then cause further symptomatology including effects on the central nervous system. Tissue destruction can be a final outcome of chemical sensitivity.

This speculative model must be subjected to several tests. Do patients with chemical sensitivities, including those with diagnosable disorders such as rheumatoid arthritis and multiple sclerosis, progress through the stages of the stress syndrome when their chemical environments are manipulated? This question can be answered by placing patients with candidate diseases in an Environmental Control Unit and studying clinical and laboratory changes in their health.

Are the mechanisms proposed above operative in chemically sensitive patients? This question can be addressed by measuring immune parameters, neurological parameters, and mediators in blood and urine while patients' chemical environments are manipulated. An Environmental Control Unit will be necessary for this type of work. Animal models may be helpful in studying the details of the response to chemical exposures depicted in figure 2.

REFERENCES

Ashford, N. A., and Miller, C. S. 1990. Multiple chemical Sensitivities. Report to the New Jersey Department of the Environment.

Ashford, N. A., and Miller, C. S. 1991. Chemical Exposures: Low Levels and high Stakes. Van Nostrand Reinhold, New York.

Biagini, R. E., W. J. Moorman, T. R. Lewis, and I. L. Bernstein. 1986. Ozone enhancement of platinum asthma in a primate model. Am. Rev. Respir. Dis. 134:719-725.

Broughton, A., and J. D. Thrasher. 1988. Antibodies and altered cell mediated immunity in formaldehyde exposed humans. Clinical Toxicology 2, 155-174, 1988.

Cochran, C. G. 1988. Immune Complex Diseases, In J. B. Wyngaarden and L. H. Smith, Jr., ed., Cecil Textbook of Medicine, 18th edition, W. B. Saunders, Philadelphia. chapter 424, p 1960.

Cullen, M. 1989. "The Worker with Multiple Chemical Sensitivities: An Overview." In, Cullen, M. (ed.) Workers with Multiple Chemical Sensitivities. Occupational Medicine: State of the Art Reviews. 2(4):655-662. Hanley & Belfus, Philadelphia.

De Sarro, G. B., Y. Masuda, C. Ascioti, M. G. Audino, G. Nistico. 1990. Behavioral and ECoG spectrum changes induced by intracerebral infusion of interferons and interleukin 2 in rats are antagonized by naloxone. Neuropharmacology 29:167-79.

Koren, H. S., R. B. Devlin, D. Hourse, S. Steingold, and D. E. Graham. The inflammatory response to the human upper airways to volatile organic compounds. 1990. Proceedings of the 5th International conference on indoor air quality and climate. Ottawa, p. 325.

Levin, A. S. & V. S. Byers. 1989. Environmental illness: a disorder of immune regulation. In M. R. Cullen, ed., Workers with multiple chemical sensitivities. State of the art reviews: Occupational Medicine. 2(4)669-681.

Meggs, W. J. "Development of a Clinical Research Protocol for the Multiple Chemical Sensitivity Syndrome."

Moore K. W., P. Vieira, D.F. Fiorentino, M. L. Trounstine, T. A. Khan, T. R. Mosmann. 1990.Homology of cytokine synthesis inhibitory factor (IL-10) to the Epstein-Barr virus gene BCRFI. Science 248:1230-4.

Obal, F. Jr., M. Opp, A. B. Cady, L Johannsen, A. E. Postlethwaite, H. M. Poppleton, J. M. Seyer, and J. M. Krueger. 1990. Interleukin 1 alpha and an interleukin 1 beta fragment are somnogenic. Amer Jour Physiology 259:R439-46.

Patterson, R. et al. 1989. IgG antibodies against formaldehyde human serum proteins: a comparison of other IgG antibodies against inhalant proteins and reactive chemicals J

Allergy Clin Immunol 84:359-366.

Randolph, T. G. 1962. Human Ecology and Susceptibility to the Chemical Environment. Charles C Thomas, Springfield, IL.

Rea, W. J., A. R. Johnson, S. Youdim, et al., 1986. T&B lymphocyte parameters measured in chemically sensitive patients and controls. Clinical Ecology 4, 11.

Riedel, , F., M. Kramer, C. Scheibenbogen, and C. H. Rieger. !988. Effects of SO2 exposure on allergic sensitization in the guinea pig. Jour Allergy Clin. Immunol. 82:527-534.

Terr, A. I. 1986. Environmental Illness: a clinical review of 50 cases. Arch Int Med 146

Considerations for the Diagnosis
of Chemical Sensitivity

William J. Rea, Alfred R. Johnson, Gerald H. Ross, Joel R. Butler,
Ervin J. Fenyves, Bertie Griffiths, and John Laseter

INTRODUCTION

The study of the effects of the environment upon the individual is now feasible due to new technology developed in the construction of environmental units.[1,2,3]. Our observations reveal that individual or multiple organs may be involved. The brain is the target organ in only a subset of chemically sensitive patients, and its involvement should not be confused with psychosomatic disease.

Over the last 16 years physicians and scientists at the Environmental Health Center in Dallas have had an opportunity to observe over 20,000 patients who had chemical sensitivity problems. These patients were studied under various degrees of environmental control. This experience is unique in the world and has resulted in numerous peer-reviewed scientific articles, chapters in books, and books on this subject.

Studies have resulted in over 32,000 challenge tests by inhalation, oral, or injection methods, of which 16,000 are double-blind. Blood chemical levels and fat biopsies for organic hydrocarbons number over 2,000, while the measurement of immune parameters are over 5,000 tests. Objective brain function tests have been accomplished in over 5,000 patients. Other objective tests, like computerized balance studies, depollutant enzyme levels, and autonomic nervous system changes as measured by the Iriscorder, number near 1,000.

We wish to share our findings with the participants of the National Academy of Sciences Committee for the study of chemical sensitivity.

DEFINITION AND PRINCIPLES

Chemical sensitivity is defined as an adverse reaction to ambient doses of toxic chemicals in our air, food, and water at levels which are generally accepted as subtoxic. Manifestation of adverse reactions depend on: (1) the tissue or organ involved; (2) the chemical and pharmacologic nature of the toxin; (3) the individual susceptibility of the exposed person (genetic make-up, nutritional state, and total load at the time of exposure); (4) the length of time of the exposure; (5) amount and variety of other body stressors (total load) and synergism at the time of reaction. (6) the derangement of metabolism that may occur from the initial insults.

To demonstrate cause-and-effect proof of environmental influence on an individual's health, one must understand several important principles and facts. These principles involve those
of *total body load* (burden), *adaptation* (masking, acute toxicological tolerance), *bipolarity*, *biochemical individuality*. Each principle will be discussed separately.

TOTAL BODY LOAD (BURDEN)

This is the patient's total pollutant load of whatever source (usually from air, food and water or surroundings [1,2,4]. The body must cope with this total burden; usually it must be utilized, expelled or compartmentalized. Total body load includes: (1) *physical factors* (e.g. hot, cold, weather changes, positive ions.[5] electromagnetic phenomena.[6] radon); (2) *toxic chemicals* (e.g. inorganics: Pb, Cd, Hg, Al, Br, etc.; organics: pesticides, formaldehydes, phenols, car exhausts, etc.).[7-21] (3) *biological* (bacteria, virus, parasites, molds,[22] food.)[23,24] (4) *psychological or emotional factors* also significantly affect the patient, confirmed by recent work in psychoneuroimmunology, linking the psyche and the neuroendocrine and immune systems.[25,26,27,28] Failure to reduce the total body load prior to pollutant challenge will frequently yield inaccurate results. Accordingly, we believe it is essential to conduct investigative procedures in controlled environmental circumstances with the total load reduced.

ADAPTATION
(MASKING, ACUTE TOXICOLOGICAL TOLERANCE)

Induced by the internal or external environment, this is a change in the homeostasis (steady state), of body function with adjustment to a new "set point".[29,30,31,32] Adaptation is an acute survival mechanism in which the individual "gets used to" a constant toxic exposure in order to survive, at the same time suffering a long-term decrease in efficient functioning and perhaps longevity. Selye was among the first to describe this compensatory mechanism.[33] Because of adaptation or tolerance, the patient's total body load may increase undetected because the *perception* of a cause-and-effect relationship is lost. With no apparent correlated symptoms, repeated exposures may continue to damage his immune and enzyme detoxification systems.[34,35] The eventual result of continued toxic exposure over a period of days, weeks, months to years is end-organ failure.[11] Withdrawal or avoidance of an offending substance for at least four days will aid in reducing the total body load, after which a controlled re-exposure challenge will reproduce cause-and-effect reactions. In these deadapted individuals, there is high reproduciblity of these evoked reactions permitting the physician to acquire sound scientific information.[36]

BIPOLARITY

After an exposure, the body initially develops a bipolar response of a stimulatory phase followed by a depressive phase,[37,29,1] usually with induction of immune and enzyme detoxification systems.[38] If the incitant is strong enough, or if substantial size or duration of exposure occurs, the induced enzyme and immune detoxification systems are depleted or depressed by overstimulation and overutilization. An individual may also initially experience a stimulatory reaction in the brain, perceiving the inciting substance not as being harmful, but as actually producing an energizing "high". Therefore, he continues to acquire more exposures. After a period of time, however, be it minutes, months, or years, his body's defenses are adversely overstimulated and he develops disabling depression-exhaustion symptoms.[31] This stimulation and depression-exhaustion pattern has been observed with many pollutant exposures, including ozone.[30,12] When studying the effects of pollutants upon adapted individuals, the stimulatory phase is often missed or misinterpreted as being normal, thus giving faulty data. Studies in the controlled environment, involving 16,000 challenges in 2,000 deadapted patients, has proven this bipolarity phenomenon repeatedly.

BIOCHEMICAL INDIVIDUALITY

Another principle necessary to understand environmental aspects of health and disease, and especially chemical sensitivity, is that of biochemical individuality. Biochemical individuality of response is the individual's uniqueness.[39] This uniqueness of response depends on the differing quantities of carbohydrates, fats, proteins, enzymes, vitamins, minerals, immune and enzyme detoxification parameters with which an individual is equipped to handle pollutant insults. These variations determine an individual's ability to process the noxious substances he encounters. They further contribute to the intensity of his reaction to toxic exposures and to his susceptibility to chemical sensitivity. Thus, a group of individuals may be exposed to the same pollutant. One person may develop arthritis, one sinusitis, one diarrhea, one cystitis, one asthma, and one may remain apparently unaffected.

We have differing quantities and interactions of carbohydrates, fats, proteins, enzymes, vitamins, minerals, and immune parameters with which to respond to environmental factors. One simple example is the noted relationship between low serum magnesium levels and the HLA B35 genotype.[40] This biochemical individuality allows us to either clear the body of noxious substances, or to collect them and contribute to the body burden. Biochemical individuality is dependent on at least three factors: genetic endowment, the state of the fetus's nutritional health and toxic body burden during pregnancy, and the individual's present toxic body burden and nutritional state at the time of exposure.

Some individuals, for example, are born with significantly lower quantities of specific enzymes (it may be 75%, 50% or even 25% of the norms). Their response to environmental stimuli is often considerably weaker than those born with 100% of the normal detoxifying enzymes and immune parameters. Examples are the babies with phenylketonuria or the individuals with transferase deficiency, who do well until exposed to their environmental triggers, and then damage sets in. There are over 2,000 genetically-transmitted metabolic errors, suggesting that most of the population will have at least one abnormality.[41] Toxic volatile organic chemicals have been shown by Laseter to bioconcentrate in the fetus, increasing the acquired burden in some babies.[42]

It is well known that some individuals acquire their toxic load at work or around their homes.[42] This changes with different seasons and weather conditions thus giving variable effects and responses over time. Extreme care must be taken in evaluation of each patient, who may exhibit unique clinical responses due to his specific biochemical individuality. As an example, it is well known that not all patients will exhibit every reported symptom associated with systemic lupus erythematosus (SLE). Similarly, each patient exposed to the same environmental pollutant will react with his or her unique complex of symptoms. Because this vital fact is misunderstood, many studies are flawed when the wrong signs and symptoms are assessed for that individual.

SPREADING PHENOMENA

Spreading is a secondary response to pollutants that can involve new incitants or new target organs. Spreading that involves new incitants occurs when the body has developed increased sensitivity to increasing numbers of biological inhalants, toxic chemicals, and foods at increasingly smaller doses. At this time, overload becomes so taxing that a minute toxic exposure of any substance may be sufficient to trigger a response or autonomous triggering my occur. For example, a person initially may be damaged by a pesticide and then eventually have his disease process triggered by exposure to a myriad of toxic chemicals and foods, such as phenol, formaldehyde, perfume, beef, lettuce, etc.

Spreading may occur for many reasons. It may be due to a failure of the detoxification mechanisms--oxidation, reduction, degradation and conjugation--brought about by toxic overloading, or it may occur because of depletion of the enzyme or coenzyme's nutrient fuels, such as zinc, magnesium, all B vitamins, amino acid, or fatty acid. This depletion may account for an increasing inability to detoxify and respond appropriately. The blood brain barrier or peripheral cellular membranes of the skin, lung, nasal mucosa, gastrointestinal or genitourinary systems may be damaged allowing previously excluded toxic and nontoxic substances to penetrate to areas that increase the risk of harm. Immune or pharmacological releasing mechanisms may become so damaged that they are triggered by many substances toxic, then non-toxic (such as food) in addition to the specific one to which they were intended to respond. It is well substantiated that antigen recognition sites may be disturbed or destroyed by pollutant overload. Hormone deregulation (feedback mechanisms) may occur allowing for still greater sensitivity.

In contrast to patients who experience increased sensitivity to multiple triggering agents, some chemically sensitive patients may have one isolated organ involved in their disease process for years, only to have dysfunction spread to other organs as their resistance mechanisms break down. This kind of spreading from one to another or multiple end-organs enables the progression of hypersensitivity and the eventual onset of fixed named disease.

SWITCH PHENOMENON

The switch phenomenon is the changing of one end-organ response to another. This usually occurs acutely, but may occur over a much longer period of time. This phenomenon was first described by Savage in the 1700s. He observed that when mental patients were at their worst, they usually had a remission of their asthma or sinusitis. When they were better mentally and they were seen in the outpatient clinic, they had a much higher incidence of sinus and asthma problems. Randolph and most other environmentally oriented physicians have also observed this phenomenon. At the EHC-Dallas, we have observed similar occurrences in our patients and, in fact, take cognizance of this phenomenon when evaluating therapy outcome.

In observing thousands of controlled challenges in the environmental unit, we have seen the target organ responses of many of our patients switch to several different ones during a long (e.g., 24-hour) reaction. Often we have seen, for example, transient brain dysfunction followed by arthralgia, followed by diarrhea, followed by arrhythmia.

Therapy can appear to have been effective, even when a pollutant has not been totally eliminated. In this case, a new set of symptoms may begin indicating that a pollutant response has simply switched to another end-organ. This phenomenon occurs frequently when symptom-suppressing medication therapy or inadequate environmental manipulation is used. For example, a patient may have his sinusitis cleared by medication (e.g., cortisone). but later, since the cause has not been eliminated, he may develop arthralgia and eventually arthritis, or his colitis may have cleared only later to have cystitis develop.

This life-long progression of disease does not have to occur if the switch phenomenon is recognized during initial evaluation. To prevent it, individuals and physicians need simply to be cognizant of seemingly unrelated events. For example, statements are often made to the effect that a child will outgrow a problem when, in reality, one symptom complex dissipates only to be replaced by a new set of symptoms. For example, a child may have recurrent ear infections. Eventually, these may stop, but bedwetting may ensue. Over time the bedwetting may cease, but the child may then develop asthma. These changes in health may appear to be totally unrelated; in this instance, however, they are switch phenomena. The situation of an adult who sprays pesticides in his home, and then visits a neurologist with complaints of headaches and a rheumatologist with symptoms of arthritis is similar. Both the physician and the patient frequently fail to recognize the relevancy of these seemingly disparate symptoms as being part of a larger pattern needing further investigation.

POLLUTION FACTS

In order to accomplish concise studies of the chemical sensitivity phenomena, one must understand some facts about environmental pollutants.

Modern technology's rapidly accelerating rate of growth has produced a wide variety of chemical products, that contribute to the total chemical environment. Recent studies show that nearly 50% of the global atmospheric pollutants are generated by man, (either isolated from natural products or synthesized), and the ubiquitous nature of the toxic chemical agents is widely appreciated.[8,13,14] It has been estimated that more than 2,000 new chemical compounds are introduced annually, and that over 60,000 different organic chemicals are used commercially today.

The widespread presence of hazardous chemicals has rendered critical the environmental sensitivity problems described by Randolph almost 40 years ago.[43] While celebrated instances of gross contamination have long been the object of professional attention, only recently have literally thousands of synthetic chemical products, heretofore believed innocuous, been incriminated as agents of homeostatic dysfunction[14,11].

Current data affirm the view that standard methods for the determination of chemical incitants may no longer be effective.[7,8,36,1] With the finding that sensitivities can occur from subthreshold and picomolar quantities of chemicals, has come the discovery that standard procedures, such as skin prick or scratch tests, often fail to demonstrate positive reactions which are otherwise verifiable.

Recent literature confirms the harmful effects of chemical incitants, like formaldehyde,[44,45] phenol,[45,46] some pesticides,[7] chlorine,[47] and petroleum alcohol,[48]

Commonly encountered chemicals like glycine,[9,49] DDT, toluene and turpentine,[50,51,52] and drugs such as hydralazine have been found to induce advanced-staged disease process.[53]

A number of familiar metals have also been incriminated, among them nickel, cobalt, chromium,[54] aluminum,[55] mercury,[56] and platinum.[57] Other common environmental chemical incitants include xylene,[58] various acrylates,[59] and acrylated prepolymers,[60] benzyl peroxide, carbon tetrachloride,[61] sulfates,[62] dithiocarbamates,[63] and diisocyanates.[64]

WATER POLLUTION

Water has an important role in delivering contaminant minerals, toxic organic and inorganic chemicals, particulate matter and radiation to the human organism. In developed nations, the incidence of many chronic diseases, particularly cardiovascular disease, is associated with water characteristics, like purity and mineral content.[13] Hardness, or the lack thereof, is involved in heart disease, hypertension, and stroke.[13] Among the theorized protective agents found in hard water are calcium, magnesium, vanadium, lithium, chromium and manganese.[13] Certainly, once cardiovascular pathology is induced, waters with high sodium content may be harmful. Other adverse agents include the metals cadmium, lead, copper, and zinc, which tend to be found in higher concentrations in soft water. Nitrates in water (usually from fertilizer) pose immediate threats to children under three months of age due to production of methemoglobin,[65] and sulfur can also cause reactions in susceptible patients.

City water, much of it secondhand, often contains from 100 to 10,000 times as many synthetic compounds as natural spring water.[66] This, coupled with the rapid growth in the use of synthetic chemicals, has focused concern on the chemical quality of drinking water.[13] Although microbes are important, attention is now being drawn to the microchemical contaminants. Advances in analytic chemistry has been able to reveal chemical contaminants in the parts-per-billion or parts-per-trillion range. It is a serious mistake to assume that extensive contamination of drinking water with "low" levels of synthetic pollutants is "normal." These chemicals are widespread, and we should not be lulled into assuming these contaminants are innocuous. Examination of our ground water has revealed many hundreds of toxic chemicals in these ranges.[67,68]

Many examples of water contamination exist and have been documented, including Times Beach, Missouri with winter floods flushing dioxin-contaminated oil used 20 years ago, Niagara's Love Canal area, Waterbury, Connecticut, and Middleboro, Kentucky.[69]

In many cases, deadly materials have been accumulating for years in dumps and landfills. In the United States, some 80,000 pits and toxic waste lagoons hold chemicals ranging from carbon tetrachloride to discarded mustard-gas bombs.[68] Slowly escaping from their burial sites, these leftovers directly contaminate our ground water. Polluted ground water exists at 347 of the nations 418 worst chemical dumps, and probably is occurring in the rest.[68] Laseter[7] and others[70] have shown that a virtual organic chemistry laboratory exists in most drinking water.

In the early 1980s, California, New York, New Jersey, Arizona, Nova Scotia, and Pennsylvania condemned dozens of public water supply wells due to trichloroethylene or tetrachloroethylene pollution.[71] Leaking fuel tanks contaminated nine Kansas public water supplies in 1981.[71] Officials in New Mexico identified 25 cities where hydrocarbons and solvents contaminated the ground water.[71] Analysis of New Orleans drinking water alone revealed the presence of 13 halogenated hydrocarbons.

Sources of water pollution fall into three major categories: (1) municipal sewage; (2) agricultural wastes; and (3) industrial wastes. Approximately 55% of the water treated in municipal plants is from homes, while another 45% is from industry. Agricultural wastes include those from livestock and toxic chemicals (pesticides, herbicides, fertilizers), and farm runoff collects in rivers, lakes, and ground water. Industrial wastes, however, contain some of our more toxic substances. Over one-half of the total volume of industrial wastes come from paper mills, organic chemical manufacturing plants, petroleum companies, and steel manufacturing. The major pollutants are chemical byproducts, oil, grease, radioactive waste and heat. Other sources of contamination are drinking water disinfectants and byproducts;[68] it should be remembered that chlorine, interacting with organic material, produces toxic trihalomethanes and other organochlorines. Alternatives to treating water with chlorine include ozone, chloramines, ultra-violet irradiation, iodination, or home reverse-osmosis and charcoal filtration.[68]

Chloride, added at many sewage treatment plants, can also react with organic matter in the water to form chlorinated hydrocarbons, many of which are also known to cause cancer. Copper sulfate, aluminum sulfate and fluorine are other major contaminants which may add to the total body burden.[68]

Over a thousand different toxic chemicals have been found in public water supplies including pesticides, herbicides, industrial solvents, and polychlorinated biphenyls, just to name a few.

Inorganic pollutants include arsenic, cadmium, chromium, copper, manganese, mercury, silver, and selenium.[13] The use of inorganic arsenic insecticides has lead to high arsenic levels in some water supplies.[13] Barium (greater than 1 mg/L) has toxic effects on the heart, blood vessels, and nerves,[68] while cadmium at levels greater than .01mg/L has adverse arterial effects. At levels greater than 1mg/L or one ppm, the following metals found as water contaminants have produced severe chronic toxicity: antimony,[72] beryllium,[73] cobalt,[73] gold,[73] iodine,[73] lithium,[73] mercury,[74] and vanadium.[73] In Minamata, Japan, between 1953 and 1960, various plastic companies dumped methyl mercury chloride into the water, producing 50 to 85 ppm of mercury in nearby fish. Four hundred and six people died after ingesting these mercury-contaminated fish, and the adverse toxicological effects on developing children are continuing to be measured.[75]

A recently completed study found that skin absorption contributed from 29 to 91% of the total body dose of pollutants (from water), with an average of about 64%.[76] This is even more important when one looks at the large number of volatile organic compounds found in our drinking and bath water.

Radiation occurs in some waters in the form of radon, a naturally occurring radionuclide that seeps from rocks and may be concentrated in airtight homes, especially the basements. At this stage, more information is needed to fully assess its effects. It probably, however, can increase the total body load.

In 1965, a serious drinking-water problem was seen in 40 percent of patients hospitalized for a program of comprehensive environmental control.[1,77,78] Today it is up to 80%. We have found that patients susceptible to water contaminants virtually always exhibit multiple sensitivities, with advanced and severe environmental reactions, especially to airborne chemicals.[1] Interestingly, water sensitivity in children was found to increase on a circadian and seasonal basis.[79] Increased severity was seen during June and July or in September and October, when grass, pollen, and mold counts were also high.[79] Some ECU patients had difficulty with waters containing high levels of sodium, others with calcium, and still others with high bicarbonate waters. A few individuals tolerated distilled water, even though it may contain some hydrocarbon residuals. Hundreds of outpatients have shown symptoms in reaction to both chlorinated and nonchlorinated waters, including numerous spring, charcoal-filtered, and distilled waters. If these water-induced symptoms remain undiscovered, food and chemical testing, may be distorted. It is vital to test and find safe water before proceeding with other testing in these severely sensitive individuals.

CHEMICAL CONTAMINATION IN FOODS

The contamination of our urban food supplies is the result of widespread use of food additives, preservatives, and dyes in growth, manufacturing and processing. Virtually all commercially grown and prepared foods have pesticides and herbicides in them.[79,9]

The literature abounds with reports of chemical sensitivities to many additives.[80,81] Contaminant reactions complicate the study of food sensitivity, forcing one to define more clearly the nature of the incitant, not only as it is encountered in foods, but in the air and water as well. Bell has reported urticarial reactions and immunological changes to exposures to a number of food additives.[82] Condemi[83] and Bell both suggest that food dyes may trigger reactions in sensitive individuals; including conditions commonly thought to be psychogenic, or certain forms of hyperactivity.[28,84,85,86,87,88,89] Lindemayer has associated urticarial reactions with several additives such as p-hydroxybenzoic acid propylester, benzoic acid, sodium benzoate, ponceau rouge, and indigo carmine.[90] Monroe's data indicate a casual role played by tartrazine azo dyes and salicylates in the provocation of vascular alterations.[36] Other additives, including sodium nitrite and sodium glutamate, have been found to trigger migraine phenomena in susceptible patients.[91]

Sulfur dioxide[16] and sodium salicylate can provoke asthmatic reactions,[92] while aspirin-like food contaminants and dyes may trigger urticaria, angioedema, bronchoconstriction and purpura.[93] An even wider variety of symptoms, including severe gastro-intestinal disorders, has been associated with sensitivities to aniline, commonly found in rapeseed oil.[94]

In our experience, natural toxic components of foods, such as alkaloids, phenols, lectins, etc. must also be accounted for when studying the secondary food sensitivity which occurs from pollutant overload in the chemically sensitive. Therefore, three factors must be considered when evaluating the total food load. These are man-made pollutant contamination, natural toxic effects of foods, and food sensitivity. Failure to consider all three in the chemically sensitive patient may color or negate otherwise a clearly defined case of chemical sensitivity.

CHEMICAL INCITANTS IN THE HOME ENVIRONMENT

Indoor air pollution in the home environment has produced a multitude of sensitivities to chemicals.[8,95] Time and space limitations allow only a cursory review of the numerous commercial hygienic products which can be noxious for chemically susceptible individuals. Among these are a wide variety of cosmetics.[96,97] particularly those containing glycerin, propylene glycol, or butylene glycol,[98] perfumes,[99] and hair products such as dyes,[100,101] creams,[102] sprays,[103] and shampoo.[104] Moreover, sensitivities have been demonstrated to occur in association with lip salve,[105] fingernail preparations,[106] soaps,[107] sanitary napkins,[108] mouthwash,[109] antiperspirants,[110] contact lenses.[111] contact lens solutions,[112] and suntan lotions.[113]

Reports are widespread of sensitivities to chemicals in textiles, including synthetic acrylic fibers,[114] polyester spin finishes,[115] the epoxy resins, and synthetic clothing.[116] Products such as fabric spray starch may also be considered toxic for the chemically sensitive individual[117] for whom even the metallic buttons on blue jeans may trigger reactions to nickel.[118] Formaldehyde[44] on synthetics or tetrachloroethylene, from dry-cleaned clothing can also produce problems.

Household cleaning products, particularly those containing formaldehyde, phenols and chlorine are hazardous for many patients. Several laundry products and detergents may be identified as household incitants[119], as well as a number of products used to clean and polish furniture[120].

The very construction of many homes may prove dangerous for the chemically sensitive patient. Data suggests that chemicals contained in wood preservatives (e.g., pentachlorophenols) are environmental incitants capable of triggering a variety of symptoms.[121,122,123] Others report problems with reactions to formaldehyde-containing pressboard, carpets, plywood and petrochemical contaminants.[124]

Current data confirm earlier findings regarding the hazards of pesticides[125] such as 2,4,DNP and fungicides[126]. Moreover, research increasingly suggests the possibility of sensitivities to apparently innocuous items such as rubber bands,[127] coins,[128] epoxy,[129] and countless paper products.[130,131] Pesticides, along with oil, gas or coal are major offenders for sensitive individuals.

Research shows house plants[132,133] and common insects[134] can now be viewed as environmental incitants or causes of homeostatic dysfunction. In addition, sensitivities to cold and heat,[36] and to contaminants in household water supplies have been associated with symptoms ranging from urticaria to severe respiratory distress.

Natural gas heat and stoves, and routine insecticiding or termite proofing of homes can be prime offenders in chemical sensitivity. One must consider these potential sources of contaminants when developing studies on chemical sensitivity. In our experience, failure to evaluate building and home environments before challenge testing will often make challenge studies invalid for the diagnosis of chemical sensitivity.

MECHANISMS

The mechanisms involved in chemical sensitivity are becoming clearer, one of which has pollutant injury occurring to the lungs or liver, with resultant free radical generation[135]. Disturbances then occurs at the cellular, subcellular and molecular levels, producing injury either immunologically, or nonimmunologically through enzyme detoxification systems. Vascular or autonomic nervous system dysfunction will then occur with one or a myriad of end-organ responses.

IMMUNOLOGICAL

Type I hypersensitivity is usually mediated through the IgE mechanism on the vessel wall. Classic examples are angioedema urticaria, and anaphylaxis due to sensitivity to pollen, dust, mold, or food,[136] or some chemicals such as toluene diisocyanate. Ten percent of the patients with immunological involvement with chemical sensitivity seen at the ECH-Dallas seem to fall within this category.

Type II cytotoxic damages may occur with direct injury to the cell. A clinical example of this is seen in patients exposed to mercury[137]. A group in Minimata, Japan developed neurological disease from eating fish exposed to toxic methyl mercury chloride. Mercurial pesticides fall into this category. Twenty percent of the patients with immunological involvement seen at the EHC-Dallas seem to fall into this Type II category.

Type III shows immune complexes of complement and gamma globulin damaging the vessel wall. A clinical example of this is lupus vasculitis. Numerous chemicals, including procainamide and chlorothaizide, are known to trigger the autoantibody reaction of lupus-like reactions. Many other toxic chemicals can also trigger the autoimmune response[138]. Other chemicals, such as vinyl chloride, will produce microaneurysms of small digital arterioles, probably due to this mechanism[139,51].

Type IV (cell-mediated) immunity occurs with triggering of the T-lymphocyte. Numerous chemicals such as phenol, pesticides, organohalides, and some metals will also alter immune responses, triggering lymphokines, and producing the Type IV reactions[138]. Clinical examples are polyarteritis nodosa, hypersensitivity angiitis, Henoch-Schonlein purpura, and Wegener's granulomatosis [1,139]. A recent study done at the Environmental Health Center - Dallas on 104 proven chemically-sensitive individuals (70 vascular, 27 asthmatic, and 7 rheumatoid), comparing them with 60 normal controls, showed that those manifesting a chemical sensitivity through their vascular tree had suppression of the suppressor T-cells (greater than 4 S.D.)[47]. Clearly the larger portion of our patients with immunological involvement fall into the Type III and IV categories.

NON-IMMUNE ENZYME DETOXIFICATION

Non-immune triggering of the cell and vessel wall may occur. Complement may be triggered directly by molds, foods, or toxic chemicals[139], and mediators like kinins, postaglandins, etc. may also be directly triggered. These reactions then cause vascular spasm, with resultant hypoxia and release of lysozymes, which further produces more spasm, hypoxia, etc. Eventually end organ failure will occur.

Triggering of the enzyme detoxification, mostly in the systems liver and respiratory mucosa, plays an important role in clearing of pollutants. It occurs, however, to a lesser extent in all systems. Foreign compound biotransformations have considerable variability, depending on genetic factors, age, sex, nutrition, health status, and the size of the dose.

The metabolism of foreign compounds usually occurs in the microsomal fraction (smooth endoplasmic reticulum) of liver cells. A few biotransformations are non-microsomal (redox reactions involving alcohols, aldehydes and ketones). There are basically four biotransformation categories -- oxidation, reduction, degradation, and conjugation.

The first three biotransformation pathways for xenobiotics are the same pathways that the body uses to process food and nutrients. If these enzyme systems are over-utilized by competing foreign pollutants, inadequate handling of food proteins can result, with the subsequent induction of food sensitivities. However, because these detoxification pathways are dependant on nutrient and mineral cofactors, these systems are *inducible* by appropriate oral or systemic supplementation. Such supplementation serves as an important factor in stabilizing and treating patients with chemical sensitivity. The fourth category of biotransformation, that of conjugation, is almost exclusively for handling foreign compounds. Conjugation appears to be uniquely utilized for the catabolism of foreign compounds, using amino acids and their derivatives with peptide bonds, and carbohydrates and their derivatives with glucide bonds. Simpler compounds like sulfate and acetate are also involved in conjugation with linkage of ester bonds. Activated conjugated compounds plus specific enzymes are often detoxified by coupling with co-enzymes, Examples: co-enzyme A with acetate, and other short-chain fatty acids adenosine or phosphoadenosine phosphate is detoxified with a methyl group from sulfate methionine, or the ethyl group from ethionine. Similarly, uridine and phosphate with glucose and glucuronic acids[140,141].

There are generally five major categories of foreign-compound conjugative processes[140]. These are: 1) *acetylation* through co-enzyme A, for detoxifying aromatic amines and sulfur amides; 2) *peptide conjugation* with glycine and aromatic carboxylic acids to hippuric acid; 3) *sulfonation* with glutathione (containing cysteine) or PAPS, and microsomal enzyme conjugation for multi-ring systems such as naphthalene, anthracene, and pheno-anthracene, which eventually results in benign mercaptic acids or alternatively benign sulfate esters; 4) *alkylations* by methionine of amines, phenols, thiols, noradrenalin, histamine, serotonin, pyridine, pyrogallol, ethylmucaptin sulfites, selenites and tellurites; 5) *Glucuronation*. Glucuronides detoxify pesticides, alcohols, phenols, enols, carboxylic acid, amino hydroxamines, carbamides, sulfonamide and thiol[140,141]. All of these processes are dependent upon nutrient fuels to keep these processes running efficiently. Toxic chemicals disturb the supply of the nutrient fuels by 1) producing poor quality food, 2) reducing intake, 3) reducing normal absorption, 4) setting up competitive absorption in the gut with nutrients, 5) imbalancing intestinal flora, 6) disturbing transport mechanisms, 7) disturbing proper decomposition and metabolism, 8) causing renal leaks, and 9) directly damaging nutrients. If nutrient inadequacy occurs, normal metabolism is overloaded and disturbed, resulting in selective changes in the pools of nutrients such as vitamins, minerals, amino acids, enzymes, lipids, and carbohydrates. Once this occurs, there is a viscous cycle of dysmetabolism, often with production or worsening of chemical sensitivity. These detoxification and metabolic defects are often measurable and have been accomplished in over 2,000 chemically sensitive patients.

DIAGNOSIS

The diagnosis of chemical sensitivity can now be made with a combination of the following history, physical examination, immune tests including IgE, IgG, complements, T & B lymphocyte subsets, blood levels of pesticides, organic compounds, heavy metals (intracellular), and occasionally objective brain function tests. Antipollutant enzymes, such as superoxide dismutase, glutathione, peroxidase, and catalase have been found to be suppressed in the chemically sensitive. Vitamin deficiencies, mineral deficiencies and excess, amino acid deficiency and disturbed lipid and carbohydrate metabolism has been observed.

Challenge tests are the cornerstone of confirmatory diagnosis. These may be accomplished through oral, inhaled, or intradermal challenges. Care should be taken to rule out inhalant problems with pollen, dust, and molds. Food sensitivity occurs in approximately 80% of the people with chemical sensitivity; and must be evaluated. When diagnosing chemical sensitivity, one must investigate water contaminant sensitivities, as 90% of people with chemical sensitivity have water contaminant reactions[4]. This can be checked by placing the patient on less chemically-contaminated, charcoal filtered, distilled, or glass-bottled spring water for four days, with subsequent rechallenge of the patient's regular drinking water. This procedure will often elicit a reaction to the water pollutants in the sensitive individual.

Patients frequently know where and when the onset of their problems occurred, e.g., sudden exposure to pesticides, working around printing machines and factory machines, etc. They usually develop increased odor perception to gasoline, perfumes, new paints, car exhausts, gas stoves, fabrics, clothing or carpeting stores, chlorine and chlorox, and cigarette smoke. Not only will they find these smells offensive, but may have marked reactions to them as well. Other symptoms can range from the almost universally-seen fatigue, to classic end-organ failures. Physical findings frequently are vascular in nature, with edema, petechiae, spontaneous bruising, purpura, or peripheral arterial spasm. Frequently there is flushing, adult-onset acne, and a yellowness of the skin without jaundice. Chronic, recurring nonspecific inflammation is usually a significant sign, e.g., colitis, cystitis, vasculitis, etc. Laboratory findings are often non-specific, e.g., sedimentation rates may increase or liver profile may be mildly off. Fifteen percent of environmentally sensitive patients have positive C-reactive proteins. Twenty-five percent show abnormal serum complement parameters. Fifty percent of the chemically sensitive patients have depressed T cells. Twenty-five percent have impaired blastogenesis, and twenty-five percent have impaired delayed hypersensitivity, as evidenced by cell-mediated immunity skin tests. Of the patients with T-cell abnormalities, the depletion of the suppressor cells is seen, by over four standard deviations from a control group of normals[141]. Ten percent of these patients have elevated IgE or IgG. Patients with recurring infections have impaired phagocytosis and killing capacity. Very accurate blood measurements are now available for the chlorinated pesticides as well. The following were found in over 200 chemically sensitive patients:

PESTICIDE IN BLOOD	% DISTRIBUTION IN 200 PATIENTS
DDT and DDE	62.0%
Hexachlorobenzene	57.5
Heptachlor Epoxide	54.0
beta-BHC	34.0
Endosulfan I	34.0
Dieldrin	24.0
gamma-Chlordane	20.0
Heptachlor	12.5
gamma-BHC (Lindane)	9.0
Endrin	5.5
delta-BHC	4.0
alpha-BHE	3.5
Mirex	2.0
Endosulfan II	1.5

Organophosphate levels are only positive within 24 hours after exposure, and are not much help. Lab tests for pentachlorophenols and organic solvents like hexane and pentane, are also now available, as are herbicide levels. General volatile organic hydrocarbons are found in a large portion of chemically sensitive patients. Their presence indicates either recent exposure, or a failure in the enzyme detoxification system. Those found in over 500 chemically sensitive patients include benzene, toluene, trimethylbenzene, xylene, styrenes, ethylbenzene, chloroform dichloromethane, 1,1,1,-trichloroethane, trichloroethylene, tetrachloroethylene, dichlorobenzenes. Metals including lead, mercury, cadmium, and aluminum are sometimes found in the intracellular contents of some chemically sensitive patients. These again are found in 10% of the patients.

Fat biopsies have been preferred on many patients with over 100 different compounds studied. Often there is more in the fat than blood in some cases such as organochlorine insecticides and more in the blood than fat such as seen with such substances as 2-methylpentane and 3-methylpentane.

Skin biopsies of bruising and petechiae reveal perivascular lymphocyte infiltrates around the vessel wall in chemically sensitive patients.

Challenge tests can be done by the sublingual or intradermal route. The efficiency of these tests is now well established as numerous studies, (several double-blind), have now been done[4,47,142,24,143,144,145]. These need to be done since 80% of the chemically sensitive are food sensitive. Blind intradermal challenge for chemicals can now be done with terpenes, petroleum derived ethanol, glycerine, formaldehyde, phenol, perfume, and newsprint, whereby production of symptoms will help establish the patient's chemical sensitivity.

Over 200,000 intradermal challenges of chemicals have been done under environmentally-controlled conditions at the EHC-Dallas. These are clearly reliable, especially as they meet the positive criteria of sign and symptom reproduction, wheal growth and negative placebo response.

Inhalation challenge is another method for the diagnosis of chemical sensitivity, done under varying degrees of environmentally controlled conditions. For best results, one uses an anodized aluminum and glass booth to do ambient dose challenge of any toxic chemical in a hospitalized, environmentally controlled setting. Some studies done in our center, under strictly controlled conditions in an environmental unit, showed significant findings (4 S.D.) of the chemical reactors over the controls when using less than .20 ppm formaldehyde, less than .0025 ppm phenol, less than .33 ppm chlorine, less than .50 ppm petroleum derived ethanol, less than .034 ppm of the pesticide, 2,4,DNP, along with 3 placebos. These tests have been used in over 3,000 patients with over 99% accuracy. Similar studies can be done in the office setting, although controls are much more difficult and one finds many more placebo reactions. This is because environmentally-controlled conditions are generally much more difficult to achieve and patients are often studied in the masked or adapted state, wherein symptoms may not be perceived. With the inhaled challenges, one can measure and plot blood levels, immune parameters, metabolic changes as well as sign and symptom scores.

Vitamin and intracellular mineral levels are needed to completely evaluate the chemically-sensitive individual. In our Center, analysis of over 300 chemically sensitive patients showed the following vitamin deficiencies: 64% with B6 deficiency, 30% with B2, 29% with B1, 27% with folic acid, 24% with vitamin D, 19% with B3, 6% with vitamin C, 3% with vitamin B12. Out of 190 chemically sensitive patients with mineral deficiencies, 88% had chromium deficiency, 12% selenium, 8% zinc, 40% magnesium and 35% sulfur. Many had mineral excess in their blood cells.

TREATMENT

The cornerstone of treatment for chemical sensitivity is avoidance[146]. This will decrease total body burden, allowing recovery of the overtaxed detoxification systems. Less chemically contaminated water (including spring, distilled, and charcoal filtered), may be used, but only in glass or steel containers. Water will leach a variety of contaminants from the walls of synthetic plastic containers. A rotary diet of less chemically contaminated food, should also be used to reduce load and keep the patient in the unmasked state. Remove as many household incitants as possible, including petroleum-derived heat, insecticides, synthetic carpets and mattresses, and formaldehyde-containing substances such as pressboard and plywood. Toxic exposures can be monitored by the general volatile organic hydrocarbon blood tests. Some job changes may be needed, while occasionally the most severely affected patients have to leave badly polluted areas. Techniques should be developed for follow-up and monitoring of these modalities.

Injection therapy for inhalants, foods, and some chemicals will also help this problem[24,144,145,147,148,149,150]. Low-dose sublingual therapy in patients with allergic rhinitis was effective[151]. These treatments can be done daily, but usually every four to seven days. In our opinion, a properly balanced rotary diet is essential in treating the patient with food sensitivity, whether or not it may be induced by chemical overload. Vitamin and mineral supplementation is often necessary to replace the deficiencies that occur from the direct toxic damage, exhausted enzymatic detoxification pathways, and from the direct competition absorption. In rare cases, nutritional replacement with intravenous hyperalimentation is needed for severely debilitated patients. Techniques should be developed for monitoring and evaluating the outcome.

CONCLUSION

The philosophy and techniques of environmental medicine developed over the last 25 years offers a means to scientifically investigate and treat patients affected by pollutants. This approach gives the physician valuable, accurate information, in the pursuit of optimum health for these environmentally-sensitive patients.

REFERENCES

1. Randolph, T.G., Human Ecology and Susceptibility to the Chemical Environment. Springfield, Illinois: Charles C. Thomas, 1962 (sixth printing 1978).

2. Dickey, L.D. (Ed.), Clinical Ecology. Springfield, Illinois: Charles C. Thomas, 1976.

3. Bell, I.R., Clinical Ecology. Bolinas, California: Common Knowledge Press, 1982.

4. Rea, W.J. and Mitchell, M.J. Chemical sensitivity and the environment. **Immun Allerg Prac**, Sept/Oct: 21-31, 1982.

5. Wordan, J.I. The effect of air in concentrations and polarity on the CO2 capacity of mammalium plasma. Fed Proc 1954; 13:557.

6. Choy, R.V.S., Monro, J.A., Smith, C.W., Electrical Sensitivities in Allergy Patients. Clin Ecol Vol IV
    No. 3. pp. 93-102.

7. Laseter, J.L.; DeLeon, I.R.; Rea, W.J.; Butler, J.R. Chlorinated hydrocarbon pesticides in environmentally sensitive patients. Clin Ecol, 2(1):3-12, 1983.

8. National Research Council. Indoor Pollutants. Washington, D.C.: Natl. Academy Press, 1981.

9. Gilpin, A. Air Pollution, 2nd Ed. St. Lucia, Queensland: University of Queensland Press, 1978.

10. Wark, K. and Warner, C. Air Pollution: Its Origin and Control. New York: Harper & Row, 1961.

11. Adelman, R.C. Loss of adaptive mechanisms during aging. Fed. Proc. 38:1968-1971, 1979.

12. Stokinger, H.E. Ozone toxicology: A review of research and industrial experience, 1954-1962. Arch Environ Health, 110:719, 1965.

13. The National Research Council. Water hardness and health. In Drinking Water and Health. New York. National Academy of Science, 1977, 439-447.

14. Winslow, S.G. The Effects of Environmental Chemicals on the Immune System: Selected Bibliography With Abstracts. Oak Ridge, TN: Toxicology Information Response Center, Oak Ridge National Laboratory, 1981.

15. Calabrese, E.J. Pollutants and high-risk groups. The Biological Basis of Increased Human Susceptibility to Environmental and Occupational Pollutants. NY: Wiley -

Interscience, 1977.

16. Freedman, B.J. Sulphur dioxide in foods and beverages: its use as a preservative and its effect on asthma. Br J Dis Chest, 74(2):128-134, 1980.

17. Davidson, M. and Fienleib, M. Disulfide poisoning: A review. Am Heart J, 83:100, 1972.

18. Innami, S., Tojo, H., Utsuja, S.; Nakamura, A.; and Nagazama,S. PCB Toxicity and Nutrition I. PCB Toxicity and Vitamin A. Jap J Nutr, 32:58-666, 1974.

19. Villeneune, D.C., Grant, D.L., Phillipos, W.E.J.; Clark, M. L.; and Clegg, D. J. Effects of PCB administration on microsomal enzyme activity in pregnant rabbits. Bull Environ Contam Toxicol, 6:120, 1971.

20. Rea, W.J.; Johnson, A.R.; Smiley, R.E.; Maynard, B.; and Dawkins-Brown, 0. Magnesium deficiency in patients with chemical sensitivity. Clinical Ecology, 4(1): 17-20, 1986.

21. Pan,Y., Johnson, A.J., Rea, W.J.: Aliphatic Hydrocarbon Solvents in Chemically Sensitive Patients. Clin Ecol 5(3):126-131,1987-88.

22. Olson, K.R., et al: Amanita phalloides-type mushroom poisoning. West. J. Med. (37:282, 1982).

23. Brostoff, Jonathan, and Challacombe Stephen; Food Allergy and intolerance, London, Bailliere Tindall/W.B. Saunders, 1987.

24. Rapp, O.J. Double-blind confirmation and treatment of milk sensitivity. Med J Aust. 1:571-572, 1978.

25. Rea, W.J.; Pan,Y. et al: Toxic Volatile Organic Hydrocarbons in Chemically Sensitive Patients, Clin Ecol 5(2):70-74,1987.

26. Hill, Johanna   Peptides and Their Receptors as the Biochemicals of Emotion, proceedings of the American Acad. Envir. Med. Annual Mtg., Nashville, TN, Oct. 29, 1987.

27. Ader, Robert Behavioral Influences of Immunity, proceedings of Amer. Acad. Envir. Med. Annual Mtg., Nashville, TN, Oct. 29, 1987.

28. King, D.S. Can allergic exposure provoke psychological symptoms? A double-blind test. Biol Psychiat. 16:3-19, 1981.

29. Randolph, T.G. Specific Adaptation. Ann Allergy. 40:333-45, 1978.

30. Mustafa, M.G. and Tierney, D.F. Biochemical and metabolic changes in the lung with oxygen, ozone, and nitrogen dioxide toxicity. Am Rev Resp Dis, 118(6), 1978.

31. Stokinger, H.E. and Coffin, D.L. Biological effects of air pollutants. In A.C. Stern (Ed.), Air Pollution. New York:
American Press, 1968.

32. Rinkel, H.J.; Randolph, T.G.; and Zeller, M. Food Allergy. Norwalk, Connecticut: New England Foundation of Allergic an Environmental Disease, reprint of 1951 text.

33. Selye, H. The general adaptation syndrome and the diseases of adaptation. J. Allergy, 17:23, 1946.

34. Jollow, D.J., et al. 1977 Biological Reactive Intermediates: Formation, Toxicity, and Innetivation. New York.

35. Wand, A.M., et al. 1976 Immunological mechanisms in the pathogenesis of vinyl chloride disease. Brit. Med. J.l, 6015:936-938.

36. Rea, W.J.; Peters, D.W.; Smiley, R.E. et al. Recurrent environmentally triggered thrombophlebitis. Ann Allergy, 47:338-344, 1981.

37. Savage, G.H. Insanity and Allied Neuroses. Philadelphia: Lea and Febiger, 1884.

38. Rowe, A.H. Allergic toxemia and migraine due to food allergy. Calif West Med. 33:785-792, 1930.

39. Williams, R.J. Biochemical Individuality. New York: Wiley, 1963.

40. Noordhout, B., et al. "Latent Tetany, magnesium and HLA tissue antigens." Magnesium-Bulletin 3 (offizielles organ Gesellshaft fur magnesium forsehung) 9, 3, 1987.

41. Stanbury, J.B., et al. The Metabolic Basis of Inherited Disease, 5th printing New York, McGraw-Hill Book Co., 1983.

42. Nadal,A. and Lee, L.Y. Airway hyperirritability induced ozone. In S.D. Lee (Ed.), Biochemical Effects of Environmental Pollutants. Michigan: Ann Arbor Science Publishers, 1977.

43. Randolph, T.G. Sensitivity to petroleum. Including the derivatives and antecedents. a. Lab Clin Med 40:931-932 Dec. 1952.

44. Fisher, S.A.: Dermatitis due to the presence of formaldehyde in certain sodium lauryl sulfate (SLS) solutions. Curtis 27(4):360-2, 366, Apr, 1981.

45. Fregert, S.: Irritant dermatitis from phenol-formaldehyde resin powder. Contact Dermatitis 6(7):493, Dec., 1980.

46. U. S. Environmental Protection Agency: Unfinished Business. A Comparative Assessment of Environmental Problems.

47. Rea, W.J., Johnson, A.R., Youdim, S., Fenyves, E.J., Samadi, N. T & B Lymphocytes. Parameters Measured in Chemically Sensitive Patients and Controls.

48. Mohave, L., Moller, J: The Atmospheric Environment in Modern Danish Dewllings -- Measurements in 39 Flats. Institute of Hygiene, University of Aarhus, Denmark. Paper presented at International Indoor Climate Symposium. Copenhagen, 1978.

49. Suhonen, R.: Contact allergy to dodecyl-di-(aminoethyl) glycine (Dedsimex i). Contact Dermatitis 6(4):290-1, Jun., 1980.

50. McMillian, R.: Environmentally-induced thrombocytopenic purpura, JAMA 242(22):2434-5, Nov. 30, 1979.

51. Rea, W.J.; Bell, I.R.; and Smiley, R.E., Environmentally triggered large-vessel vasculitis. In F. Johnson and J.T. Spence (Eds.), Allergy: Immunology and Medical Treatment. Chicago: Symposia Specialist, 1975.

52. Rea, W.J. Review of cardiovascular disease in allergy. In C.A. Frazier (Ed.), Biannual Review of Allergy. Springfield, Illinois, 1979-80.

53. Peacock, Al: Case reports, hydralazine-induced necrotizing vasculitis, Br Med J 282(6270):1121-2, April 4, 1981.

54. Dooms-Goossens, A., Ceuterick, A., Vanmaele, N., Degreef, H.: Follow-up study of patients with contact dermatitis caused by chromates, nickel, and cobalt. Dermatologica 160(4):249-60, 1980.

55. Clemmensen, 0., Knudsen, H.E.: Contact sensitivity to alumi- num in a patient hyposensitized with aluminum precipitated grass pollen. Contact Dermatitis 6(2):305-8, Aug., 1980.

56. Vermeiden, Oranje, A.P., Vuzevski, V.0.: Stolz, E. Mercury exanthem as occupational dermatitis. Contact Dermatitis, 6(2):88-90, Jan., 1980.

57. Hughes, E.G.: Medical surveillance of platinum refinery workers. Soc Occup Med 30(1):27-30, Jan., 1980.

58. Bell, I., King, D.: Psychological and physiological research relevant to clinical ecology: an overview of the recent literature, Clin Ecol, 1982.

59. Bjorkner, B., Dahlquist, I., Fregert,S.: Allergic contact dermatitis from acrylates in ultraviolet curing inks. Contact Dermatitis 6(6):405-9, Oct. 1980.

60. Bjorkner, B.: Sensitization capacity of acrylated prepolymers in ultraviolet curing inks tested in the guinea pig. Acta Derm Venercol (Stockh), 61(1):7-10, 1981.

61. Romaguera, Gimalt, Sensitization to benzoyl peroxide, retinoic acid and carbon

tetrachloride. Contact Dermatitis, 6(6):442, Oct., 1980.

62. Satish, J. Importance and measurement of sulfates in floating dust. Soz Praeventivmed 25(4):201-2, Sep., 1980.

63. Kleibl, K., Rackova, M.: Cutaneous allergic reactions to dithiocarbamates. Contact Dermatitis 6(5):348-9, Aug., 1980.

64. Gallagher, J.S., Tse, C.S., Brooks, S.M., Bernsten, I.I.: Diverse profiles of immunoreactivity in toluene diisocyanate (TOI) asthma. JOM 23(9):610-6, Sep., 1981.

65. Wilson C.W.M. Hyperactivity to Maine tap water in children: its clinical features and treatment. Nutri and Health 2: 51-63, 1983.

66. Glaze, W.E., North Texas State University, Environmental Division, Denton, Texas, Personal communication.

67. Cherry, R. and Cherry, F.: What's in the water we drink? The New York Times Magazine, Dec. 8, 1974.

68. Environmental Protection Agency (EPA): Is your drinking water safe? (Booklet)

69. Trimble, J. U.S. News and World Report, Feb. 28, 1984, pp.27-32.

70. Spalding, R.F.; Junk, G.A. Richard, J.J. Water: Pesticides in ground water beneath irrigated farmland in Nebraska, Pesticides Monitoring Journal 1(2): 70-73, 1980.

71. Junk, G.A.; Spalding, R.F. and Richard, J.J. Aureal, Vertical, and temporal differences in ground water chemistry. II. Organic constituents, 1983.

72. Corwin, A.H. Heavy metals in air, water, and our habitat In Dickey, L.D. (Ed) Clinical Ecology, Springfield, Il: Charles C. Thomas, 1976, pp. 292-305.

73. Moran, J.M.; Morgan, M.O. Wiersma J.N Introduction to Environmental Science 2nd Ed. New York: W.H. Freeman and company, 1986, pp. 177-209.

74. Skiba, N.S. Role of river discharge of solids in migration of mercury, Zap Kirgizsk Otd Vses Mineralog Obschchestva 4:31, 1963; Chem Abst, 61:5357e, 1964.

75. Takenchi, T. Morikawa, N.; Matsumoto, H.; and Shiraishi, Y. Apathological study of Minamata disease in Japan, Acta Neuropathol 2:40, 1962.

76. Brown, H.S. Bishop, D.R.; Rowan, C.A. The role of skin absorption as a route of exposure for volatile organic compounds in drinking water, Am J Pub Health 74: 479-484, 1984.

77. Randolph, T.G. Ecologic orientation in medicine: Comprehensive environmental control in diagnosis and therapy. Ann Allergy. 23:7-22, 1965.

78. Randolph, T.G. The ecologic unit. Hospital Management, Part I, 97:45, 1965; Part II, 97:46, 1965.

79. Wilson, C.W.M. Hypersensitivity to Maine tap water in children: its clinical features and treatment. Nutrition and Health 2:51-63, 1983.

80. Juhlin, L.: Incidence of intolerance to food additives. Int J Dermatol 19(10):548-51, Dec., 1980.

81. Fries, J.H.: Food allergy: Current concerns. Ann Allergy 46(5):260-3, May, 1981.

82. Bell, I, King, D.: Psychological and physiological research relevant to clinical ecology: An overview of the recent literature. Clin Ecol, 1982.

83. Condemi, J.J.: Aspirin and food dye reactions.Bull N.Y.   Acad Med 57(7):600-7, Sep., 1981.

84. Rea, W.J. Environmentally triggered small vessel vasculitis. Annals of Allergy, 38:245-51, 1977.

85. Potkin, S.G.; Weinberger, D.; Kleinman, J. et al. Wheat gluten challenge in schizophrenic patients. Amer J. Psychiat.138:1208-1211, 1981.

86. Hanninen, H.; Eskelinen, L.; Husman, K.; and Nurminen, M. Behavioral effects of long-term exposure to a mixture of organic solvents. Scand. J. Work Environ. Health. 4:240-55, 1976.

87. Randolph, T.G. The history of ecologic mental illness. In C.A. Frazier (Ed.), Annual Rev Allergy 1973. Flushing, NY: Med Exam Pub Co, 1974.

88. Savage, G.H. Insanity and Allied Neuroses Philadelphia: Lea and Febiger, 1884.

89. Rea, W.J.; Butler, J.R.; Laseter, J.L.; and DeLeon, I.R. Pesticides and brain function changes in a controlled environment. Clinical Ecology, 2(3):145-50, 1984.

90. Lindemayr, H. Schmidt, J.: Intolerance to acetylsalicylic acid and food additives in patients suffering from recurrent urticaria (author's transl.). Wien Klin Wochenschr 91(24):817-22, Dec., 21, 1979.

91. Hanington, E.: Diet and migraine. J. Hum. Nutr. 34:175-80,1980.

92. Dahl, R.:   Sodium salicylate and aspirin disease. Allergy 35(2):155-6, Mar., 1980.

93. Wuthrich, B., Fabro, L.: Acetylsalicylic acid and food additive intolerance in urticaria, bronchial asthma and rhinopathy. Schweiz. Med. Wochenchr. 111(39):1445-50, Sep. 26, 1981.

94. Tabuenca, J.M.: Toxic-allergic syndrome caused by ingestion of rapeseed oil

denatured with aniline. Lancet 12:2(8246):567-8, Sep. 1981.

95. Cullen M. ed. Workers with Multiple Chemical Sensitivities. Occup. Med. State of the Art Reviews. 2(4),1987.

96. Thune, P: Photosensitivity and allergy to cosmetics. Contact Dermatitis 7(1):54-5, Jan., 1981.

97. Zaynoun, S.T., Aftimos, B.A., Tenekjian, K.K., Kurban, A.K.: Berloque dermatitis a continuing cosmetic problem. Contact Dermatitis 7(2):111-6, Mar., 1981.

98. Fisher, A.A.: Reactions to popular cosmetic humectants, Part III. Glycerin, propylene glycol, and butylene glycol. Curtis 26(3):243-4, 269, Sep., 1980.

99. Guin, J.D., Berry, V.K.: Perfume sensitivity in adult females. J Am Acad Dermatol 3(3):299-302, Sep., 1980.

100. Foussereau, J., Reuter, G., Petitjean, J.: Is hair dyed with PPD-like dyes allergenic? Contact Dermatitis 6(2):143, Jan., 1980.

101. Klein, A.D. 3d, Rodman, O.G.: Allergic contact dermatitis to paraphenylenediamine in hair dye. Milit Med 146(1):46-7, Jan., 1981.

102. Ramsay, C.A.: Skin responses to ultraviolet radiation in contact photodermatitis due to fentichlor. J Invest Dermatol 72(2):99-102, Feb., 1979.

103. Schlueter, L.E., Soto, R.J., Baretta, E.D., Herrmann, A.A., Ostrander, L.E., Stewart, R.D.: Airway response to air spray in normal subjects and subjects with hyperreactive airways. Chest 75(5):544-8, May, 1979.

104. Stoll, D., King, L.E., Jr.: Disulfiram-alcohol skin reaction to beer-containing shampoo. JAMA 244(18):2045, Nov., 1980.

105. Hindson, Phenyl salicylate (Salol) in a lip salve. Contact Dermatitis 6(3):216, Apr., 1980.

106. Fisher, A.A.: Permanent loss of finger nails from sensitization and reaction to acrylic in a preparation designed to make artificial nails. J Dermatol Surg Oncol 6(1):70-1, Jan., 1980.

107. Jordan, W.P., Jr.: Contact dermatitis from D & C yellow 11 dye in a toilet bar soap. J Am Acad Oermatol 4(5):613-4, May, 1981.

108. Larson, W.G.: Sanitary napkin dermatitis due to the perfume. Arch Dermatol 115(3):363, Mar., 1981.

109. Mathias, C.G., Chappler, R.R., Maibach, H.I.: Contact urticaria from cinnamic aldehyde. Arch Dermatol 116(1):74-6, Jan., 1980.

110. Aust, L.B., Maibach, H.I.: Incidence of human skin sensitization to isostearyl alcohol in two separate groups of panelists. Contact Dermatitis 6(4):269-71, Jun., 1980.

111. Sertoli, A., Di Fonzo, E., Spallanzani, P., Panconesi, E.: Allergic contact dermatitis from thimerosal in a soft contact lens wearer. Contact Dermatitis 6(4):292-3, Jun., 1980.

112. Zeigen, S.R., Jacobs, I.H., Weinberger, G.I.: Delayed hyper-sensitivity to thimerosal in contact lens solutions. J Med Soc NJ 78(5):362-4, May, 1981.

113. Jackson, R.T., Nesbitt, L.T., Jr., DeLeo, V.A.:6-Methylcoumarin photocontact dermatitis. J Am Acad Dermatol 2(2):124-7, Feb., 1980.

114. ElSaad El-Rifaie, M.: Perioral dermatitis with epithelioid cell granulomas in a woman: a possible new etiology. Acta Derm Venereal (Stockh) 60(4):359-60, 1980.

115. Burrows, D., Campbell, H.S.: Contact dermatitis to polyesters in finishes. Contact Dermatitis 6(5):362-3, Aug., 1980.

116. Imbeau, S.A., Reed, C.E.: Nylon stocking dermatitis. An unusual example. Contact Dermatitis 5(3):163-4, May, 1979.

117. McDaniel, W.H., Marks, J.G., Jr.: Contact urticaria due to sensitivity to spray starch. Arch Dermatol 115(5):628, May, 1979.

118. Larson, F.S., Brandrup, F.: Nickel release from metallic buttons in blue jeans. Contact Dermatitis 6(4):298, Jun, 1980.

119. Rudzki, E., Kozlowska, A.: Causes of chromate dermatitis in Poland. Contact Dermatitis (63):191-6, Apr., 1980.

120. Kleinhans, Days. Formaldehyde contact allergy. Derm Beruf Umwelt 28(4):101-3, 1980.

121. Klaschka, F., von Nieding,F., Walter, R. Skin sensitization to wood preservatives containing pentachlorophenol (PGP). Zentralbl Arbeitsmed Arbeitsschutz Prophyl 29(6):150-4, Jun., 1979.

122. Bach, B., Pedersen, N.B.: Contact dermatitis from a wood preservative containing tetrachloroisophthalonitriles. Contact Dermatitis 6(2):142, Jan., 1980.

123. Spindeldreier, A., Deichmann, B.: Contact dermatitis against a wood preservative with a new fungicidal agent. Derm Beruf Umwelt 28(3):88-90, 1980.

124. Frigas, E., Filley, W.V., Reed, C.E.: Asthma induced by dust from urea-formaldehyde foam insulating material. Chest 79(6):706-7, Jun., 1981.

125. Matsushita, T., Nomura, S., Wakatsuki, T.: Epidemiology of contact dermatitis from

pesticides in Japan. Contact Dermatitis 6(4):255-9, Jun., 1980.

126. Rudzki, Mapiorkowska, Dermatitis caused by the Polish fungicide Sadoplon 75. Contact Dermatitis 6(4):300-1, Jun, 1980.

127. Weinstein, L.H., Fellner, M.J.: Rubber band dermatitis. Int J Dermatol 18(7):558, Sep., 1979.

128. Conde-Salazar, L., Romero, L., Guimaraens, D.: Metal content in Spanish coins. Contact Dermatitis 7(3):166, May, 1981.

129. Fregert, S.: Epoxy dermatitis from the nonworking environment. Br J Dermatol 105 Sppl 21:63-4, Sep., 1981.

130. Guin, J.D.: Contact dermatitis to perfume in paper products letter. J Am Acad Dermatol 4(6):733-4, June, 1981.

131. Marks, J.G., Jr.: Allergic contact dermatitis from carbonless copy paper. JAMA 245(22):2331-2. June 12, 1981.

132. Calnan, C.D.: Dermatitis from ivy (Hedera canariensis variegata). Contact Dermatitis 7(2):124-5, Mar., 1981.

133. Sugai, T., Takahashi, Y., Okuno, F.: Chrysanthemum dermatitis in Japan. Contact dermatitis 6(2):155, Jan., 1981.

134. Fuchs, E. Insects as inhalant allergens. Allergol Immunopathol (Madr) 7(3):227-30, May-June, 1979.

135. Levine, S. and Kidd, P. Antioxidants Adaptations, BioDivision, Allergy Research Group, San Leando, California, 1985.

136. Theorell, H.; Blambock, M; Kockum, C. Demonstration of reactivity to airborne and food antigen in cutaneous vasculitis by variation in fibrino6peptide and others, blood coagulation, fibrinolysis, and complement parameters, Thrombo Haemo Sts (Stattz), 36: 593, 1976.

137. Gaworoki, C.L. and Sharma, R.P. The effects of heavy metals on (3H) thymidine uptake in lymphocytes, Toxicol App Pharmacol, 46(2): 305-13, 1978.

138. Winslow, S.G. The effects of environmental chemicals on the immune system: a selected bibliography with abstracts, Oak Ridge, Tennessee: Toxicology Information Response Center, Oak Ridge National Laboratory, 1981, pp. 1-36.

139. Rea, W.J. and Suits, C.W. Cardiovascular disease triggered by foods and chemicals In Gerrard, J.W. (Ed.) Food allergy: new perspectives, Springfield,Illinois: 1980. Charles C. Thomas, 1980.

140. Jakoby, W.B. (Ed.) Enzymatic Basis of Detoxication, Vol. I 1983, Academic Press.

141. Reeves, A.L. (Ed.) Toxicology: Principles and Practice Vol. 1981, John Wiley & Sons, New York.

142. O'Shea, J.; Porter, A. Double-blind study of children with hyperkinetic syndrome treated with multi-allergen extract sublingually, J Learning Disabil 14: 189-91, 1981.

143. Forman, R.A. Critiques of evaluation studies of sublingual and intracutaneous provocation tests for food allergy, Med hypotheses 7: 1019-27, 1981.

144. Miller, J.B. A double-blind study of food extract injection therapy, Ann Allergy 38(3): 185-91, 1977.

145. Boris, Shiff, Weindorf, S. Bronchoprovocation blocked by neutralization therapy, J Allergy Clin Immunol 7: 92, 1983.

146. Rea. W.J., Pan,Y, Johnson,A.R. Clearing of Toic Volatile Hydrocarbons from Humans. Clin Ecol 5(4)"166-170, 1987/88.

147. Boris M., Schiff M., Weindorf S. Antigen-Induced Asthma Attenuated by Neutralization Therapy. Clin Ecol, 1985; 3:59-62.

148. Boris M, Schiff M, Weindorf S. Injection of Low-dose Antigen Attenuates the Response to Subsequent Bronchoprovocative Challenge. Otolaryngol-Head Neck Surg, 1988; 98:538-545.

149. Miller JB, Intradermal Provocative/Neutralizing Food Testing and Subcutaneous Food Extract Injection Therapy. In Brostoff J, Challacombe S, (Eds): Food allergy and Intolerance, London, Bailliere Tindall Publishers, 1987: 932-946.

150. Rea WJ, et al: Elimination of Oral Food Challenge Reaction by Injection of Food Extracts, Arch Otolaryngoll, 1984; 110:248-252.

151. Scadding G. Brostoff, J.: Low-dose Sublingual Therapy in Patient With Allergic Rhinitis Due to House Dust Mite, Clin Allergy, 1986; 16: 483-491.

Occupational Asthma

Laura S. Welch

Asthma consists of reversible airway obstruction. There is no uniform definition of occupational asthma, but most experts would call it a new state of bronchial hyper-reactivity secondary to some agent at work [1], or reversible obstruction of the airways caused by inhalation of a substance or material used by a worker or present at his work [2]. Asthma can be a pre-existing disease that is exacerbated by exposures at work, rather than having these exposures as the primary cause; this is not defined as occupational asthma but rather as exacerbation of pre-existing disease. Exacerbation of pre-existing disease is work-related, but I will reserve the term occupational asthma for use as defined above.

Three percent of all Americans have asthma [3], and 5 to 15% of all asthma is estimated to be occupational [4,5]. The talk will present identified causes of occupational asthma, and working case definitions of the disease.

Even though there has been a great deal of research, not everyone will agree on the case definition of occupational asthma. Definitions vary by the study and the instrument used. Questionnaire has been useful for screening, but are less sensitive than methacholine challenge in detecting cases [6]. If pulmonary function is used, as usual definition is a drop of greater than 10% in FEV1 over a work shift, or a greater than 20% drop in peak flow over a work day. It may take repeated measurements in any one worker to detect such a drop, for some react only at the end of a work week.

Much has been written on the mechanism of occupational asthma, and both Dr. Liebowitz and Dr. Karol have discussed this in their presentations; I will not review that in detail here. As an overview, occupational asthma can be allergic or non-allergic. Asthma may develop as a specific sensitivity to one agent (or a family of related agents), or as a response to exposure to non-specific irritants.

There is certainly a major role of the immune system in the development of occupational asthma specific to one agent. In studies of asthma secondary to Western red cedar dust, considered a prototype of asthma from low-molecular weight compounds, investigators have found specific IgE in 40% of cases (but note that it is not detectable in 60%) [4]. Studies of isocyanate-induced asthma have also found IgE in some cases but not all. Many investigators think there is a role for direct histamine release in the lung. But in addition, there is much emphasis on the role of inflammation in the initiation and continuation of asthma.

The clinical presentation of occupational asthma can consist of an immediate reaction, a late-phase reaction, or commonly a mixture of both types [2,4]. In Western red cedar cases, for example, late asthmatic reactions alone occurred in 44%, both in 49% and immediate reactions alone in only 7%. There was no correlation between development of red cedar asthma and previous atopic status. These findings suggest that the allergic reaction is more complex than IgE alone. In addition, this same study reported that nonspecific bronchial hyper-reactivity (as measured with methacholine) correlated strongly with specific reactivity to plicatic acid, and decreased as patients became asymptomatic.

The second syndrome, a response to what we think is non-specific irritation rather than a specific allergy, was described by Brooks [7], and called RADS (reactive airway dysfunction syndrome); characteristics of this group of patients was also described by Tarlo and Broder [8]. Tarlo reported that this syndrome made up 10% of patients coming for evaluation to two centers. In reviewing 15 cases, these individuals has a low incidence of atopic disease, and were more likely to be smokers. These findings need further investigation.

What we know about occupational asthma can help us in beginning research on multiple chemical sensitivity. Study of occupational asthma first and foremost illustrates both the difficulty of developing a case definition, and the important of having one, as well as the importance of standardized measurement. These are basic research concepts, but it has taken many scientists to develop what tests and clinical signs are appropriate for the study of occupational asthma.

An approach for epidemiologic study that has worked with asthma is to use a questionnaire to define cases, and then proceed to a more detailed case investigation. We know that we do not capture all occupational asthma cases with a questionnaire, but this approach is efficient, acceptable, and yields very useful information of the subgroups of asthmatics that have a set of symptoms.

Occupational asthma is one disease for which exposure is part of the case definition. A particular set of symptoms of clinical findings develops in relationship to exposure. I expect that a case definition of MCS will include such a response to exposure as well, and we can use the lessons learned from occupational asthma to guide us here.

Occupational asthma has many causative agents, and more than one mechanism that results in the same clinical presentation. In asthma, there is clearly a role for both the immune system and direct inflammation.

Of particular importance, studies of occupational asthma have taught us about the role of controlled provocative challenges, and the importance of these challenges being performed in a standard fashion in a few specialized centers [9]. The diagnosis of occupational asthma usually does not need a controlled challenge, and such studies serve as a research toll primarily.

Finally, we have learned a great deal about the role of testing for specific antibodies or other tests of the immune system. Research in occupational asthma has shown that great attention must be paid to quality assurance and quality control, and that analyses must be standardized between laboratories [10].

REFERENCES

1. Hendrick, D.J., Fabbri, L.: Compensating occupational asthma. Thorax 1981; 36: 881-884.

2. Brooks, Stuart M.: Bronchial asthma of occupational origin. Scand J Work Environ Health 1977; 3: 53-72.

3. Occupational Disease Surveillance: occupational asthma. MMWR 1990; 39: 119-123.

4. Chan-Yeung, M., Lam, S.: Occupational asthma. Am Rev Respir Dis 1986; 133: 686-703.

5. Blanc, P.D.: Occupational asthma in a national disability survey. Chest 1987; 92: 613-617.

6. Bernstein, D.I., Cohn, J.R.: Guidelines for the diagnosis and evaluation of occupational immunologic lung disease: preface. J Allergy Clin Immunol 19; 5: 791-793.

7. Brooks, Stuart M., Weiss, Mark A., Bernstein, I.L.: Reactive airways dysfunction syndrome--case reports of persistent airways hyperreactivity following high-level irritant exposures. J Occup Med 1985; 27: 473-476.

8. Tarlo, S.M., Broder, I.: Irritant-induced occupational asthma. Chest 1989; 96(2): 297-300.

9. Brooks, Stuart M.: The evaluation of occupational airways disease in the laboratory and workplace. J Allergy Clin Immunol 1982; 70: 56-66.

10. Grammer, L.C., Patterson, R., Zeiss, C.R.: Guidelines for the immunologic evaluation of occupational lung diease. J Allergy Clin Immunol 1989; 84: 805-813.

Workshop
To Develop Research Protocols for
Multiple Chemical Sensitivity

Contributors

Nicolas A. Ashford
    Massachusetts Institute of Technology

Iris R. Bell
    Department of Psychiatry, University of Arizona

Robert Burrell
    School of Medicine, West Virginia University

Joel B. Butler
    Clinical and Behavior Psychology, Dallas

Devra Lee Davis
    National Academy of Sciences

Roy L. DeHart
    University of Oklahoma Health Sciences Center

Ervin J. Fenyves
    University of Texas at Dallas

Nancy Fiedler
    Department of Environmental and Community Medicine, UMDNJ

Bertie Griffiths
    Environmental Health Center, Dallas

Gunnar Heuser
    School of Medicine, UCLA

S. Heuser
    Director, Environmental Medical Research and Information Center, Los Angeles

Alfred R. Johnson
    Environmental Health Center, Dallas

Howard Kipen
    Department of Environmental and Community Medicine, UMDNJ

John Laseter
    AccuChem Laboratories, Richardson, Texas

Michael Lebowitz
    College of Medicine, University of Arizona

Claudia S. Miller
    The University of Texas Health Science Center at San Antonio

William J. Meggs
    Department of Medicine, East Carolina University

William J. Rea
    Environmental Health Center, Dallas

Gerald H. Ross
    Environmental Health Center, Dallas

Jonathan Samet
    University of New Mexico

Laura S. Welch
    School of Medicine, George Washington University

A. Wojdani
    Director Immunosciences Laboratory, Los Angeles

Workshop
To Develop Research Protocols for
Multiple Chemical Sensitivity

Participants

Jonathan Samet, *Chairman*
Pulmonary Division
UNMH Tumor Registry
2211 Lomas N.E.
Albuquerque, NM 87131

Roy E. Albert
Department of Environmental Health
University of Cincinnati
3223 Eden Avenue
Cincinnati, OH 45267-0056

Joseph F. Albright
Basic Immunology Branch
National Institute of Allergy
and Infectious Diseases
National Institutes of Health
Room 757, Westwood Building
Bethesda, MD 20892

Nicholas A. Ashford
Associate Professor of
Technology and Policy
Massachusetts Institute of Technology
77 Massachusetts Avenue
Cambridge, MA 02139

Robert B. Axelrad
Indoor Air
Environmental Protection Agency
401 M Street S.W. (ANR-445W)
Washington, DC 20460

Rebecca Bascom
Assistant Professor of Medicine
University of Maryland
22 South Greene Street
Baltimore, MD 21201

Iris R. Bell
Department of Psychiatry
University of Arizona
Health Sciences Center
1501 N. Campbell Avenue
Tucson, AZ 85724

Eula Bingham
University of Cincinnati
College of Medicine
Kettering Lab, ML-56
Cincinnati, OH 45267

John H. Boyles, Jr.
7076 Corporate Way
Centerville, OH 45455

Robert Burrell
Department of Microbiology
West Virginia University
2082 HSN
Morgantown, WVA 26506

James E. Cone
University of California
San Francisco General Hospital
Building 9, Room 109
San Francisco, CA 94110

Jeff Davidson
Environmental Protection Agency
401 M Street SW (PM-273)
Washington, DC 20460

Earon S. Davis
Environmental Health Consultant
2530 Crawford Avenue, Room 115
Evanston, IL 60201

Roy DeHart
Division of Occupational Medicine
University of Oklahoma
School of Medicine
800 N.E. 15th Street, Room 218
Oklahoma City, OK 73104

Dan Ein
4636 Kenmore Drive, N.W.
Washington, DC 20007-1924

Nancy Fiedler
Environmental & Community Medicine
Robert Wood Johnson Medical School
675 Hoes Lane
Piscataway, NJ 08854

Jordan Fink
Allergy Department
Medical College of Wisconsin
8700 West Wisconsin Avenue
Milwaukee, WI 53226

Elizabeth Gresch
The Dow Chemical Co.
2030 Dow Center
Midland, MI 48674

Judy Guerriero
University of California
San Francisco General Hospital
Building 9, Room 109
San Francisco, CA 94110

Gunnar Heuser
NeuroMed and NeuroTox Associates
323 S. Moorpark Road
Thousand Oaks, CA 91361
805, 497-3518

William J. Hirzy
Environmental Protection Agency
401 M Street, SW (TS-778)
Washington, DC 20460

Ken Hudnell
Office of Research & Development
Environmental Protection Agency
Neurophysiology Branch
Mail Drop 74-B
Research Triangle Park, NC 27711

Pauline Johnston
Indoor Air
Environmental Protection Agency
401 M Street SW (ANR-445W)
Washington, DC 20460

Meryl Karol
Graduate School of Public Health
Environmental and Occupational Health
University of Pittsburg
130 DeSoto Street
Pittsburg, PA 15261

Howard R. Kehrl
Physiology Section
Environmental Protection Agency
Human Studies Division
Clinical Reserach Branch, HERL/ORD
Building C-224-H
Chapel Hill, NC 27514

Howard Kipen
Environmental & Community Medicine
Robert Wood Johnson Medical School
675 Hoes Lane
Piscataway, NJ 08854

Loren D. Koler
College of Veterinary Medicine
Oregon State University
Corvallis, OR 97330

Michael D. Lebowitz
Professor of Internal Medicine
University of Arizona College
of Medicine
1501 N. Campbell Avenue
Tucson, AZ 85724

Max Lum
Director, Health Education
ATSDR
1600 Clifton Road
MS E-33
Altanta, GA 30333

Roberta Madison
Department of Health Sciences
California State University
18111 Nordhoff Street
Northridge, CA 91330

William J. Meggs
Assistant Professor of Medicine
Allergy/Immunology
East Carolina University School
of Medicine
Brody Building
Greenville, NC 27858-4354

Mark Mendel
EETB, California State Department
of Health
5900 Hollis, Suite E
Emeryville, CA 94608

Claudia S. Miller
Allergy and Immunology
University of Texas
Health Science Center
7703 Floyd Curl
San Antonio, TX 78284

Frank Mitchell
Senior Medical Officer
ATSDR
1600 Clifton Road
MS E-33
Atlanta, GA 30333

Lars Molhave
John B. Pierce Foundation Laboratory
Department of Epidemiology
& Public Health
Yale University School of Medicine
290 Congress Avenue
New Haven, CT 06519

David Ozonoff
Boston University
School of Public Health
80 East Concord Street
Boston, MA 02118

Andrew Pope
Institute of Medicine
National Research Council
2101 Constitution Avenue, NW
IOM-2137
Washington, DC 20418

Patricia H. Price
ATSDR
1600 Clifton Road
MS E-33
Atlanta, GA 30333

William J. Rea
8345 Walnut Hill Lane, Suite 205
Dallas, TX 75231-4262

Kathleen E. Rodgers
Livingstone Laboratories
1312 N. Mission Road
Los Angeles, CA 90033

Anne Sassaman
Division Director, Extramural Program
DERT, NIEHS
North Campus., Building 3, Room 301A
Research Triangle Park, NC 27709

Jan Stolwijk
John B. Pierce Foundation Laboratory
Department of Epidemiology
& Public Health
Yale University School of Medicine
290 Congress Avenue
New Haven, CT 06519

Virginia Sublet
ATSDR
1600 Clifton Road
Atlanta, GA 30333

Abba I. Terr
450 Sutter Street
San Francisco, CA 94108

Jack D. Thrasher
11330 Quail Creek Road
Northridge, CA 91326

William J. Waddell
Professor and Chairman
Department of Pharmacology
& Toxicology
University of Louisville
School of Medicine
Louisville, KY 40292

Laura Welch
Associate Professor of Medicine
George Washington University
School of Medicine
2150 Pennsylvania Venue, NW
Washington, DC 20037

James E. Woods
W. E. Jamerson Professor of Building
Construction, College of Architecture
and Urban Studies
Virginia Polytechnic Institute & State
University
202 Cowgill Hall
Blacksburg, VA 24061-0205

Grace Ziem
Occupational and Environmental Health
Research Office
1722 Linden Avenue
Baltimore, MD 21217

STAFF:
Robert P. Beliles
National Research Council, Room HA
354
2101 Constitution Avenue, NW
Washington, DC 20418

Devra Lee Davis
National Research Council, Room NAS
341b
2101 Constitution Avenue, NW
Washington, DC 20418

Richard Thomas
National Research Council, Room HA
354
2101 Constitution Avenue, NW
Washington, DC 20418

Sharon Holtzmann
National Research Council, Room NAS
359
2101 Constitution Avenue, NW
Washington, DC 20418